The Reconstruction of the Sinister Universe

The Final Compilation

Greg Feild

August 26, 2022

Abstract

This volume is yet another compilation of papers concerning speculations about 'The Sinister Universe': Postmodern Physics, Mechanism and Energetics, Deconstructing Modern Physics, and Empirical Enquiries.

Postmodern Physics

Principia Quantuma Mechanicum

physicis sit amet !

Greg Feild

November 1, 2020

'Second Edition'

About the author:

Greg Feild is a physicist with a PhD from the Pennsylvania State University.

He is the author of a quantum mechanical theory of everything, and has written several books on the subject.

He is also the manservant to two very bad dogs.

The deluded, imagining trivial things to be vital to life, follow their vain fancies and never attain the highest knowledge

But the wise, knowing what is trivial and what is vital, set their thoughts on the supreme goal and attain the highest knowledge

-- [the Buddha]
The Dhammapada

Abstract:

In this book, we examine the world in the light of our new universal model of physics.

We will visit all of our favorite themes and continue the investigations into several topics still incomplete since our last book “Toward a Metaphysics of Mass and Motion”, such as; our new *physical interpretation* of spin and our new *physical model* of the only three elementary subatomic particles; the photon, the neutrino and the electron.

The universal model is field, and vacuum, free. Those paradigms are tired, and retired.

There is **only** mass in motion.

There are only atoms and the void

For I see a man must either resolve to put out nothing new
or to become a slave to defend it.

-- Isaac Newton

Greek philosophy for physicists:

And assuredly no one will argue that there is any other method [than dialectic] of comprehending by any regular process all true existence, or of ascertaining what each thing is in its own nature; for the arts in general are concerned with the desires and opinions of men, or are cultivated with a view to production and constructions; or for the preservation of such productions and constructions; and as to the mathematical sciences which, as we were saying, have some apprehension of true being -- geometry and the like -- they only dream about being, but never can they behold the waking reality so long as they leave the hypotheses which they use unexamined, and are unable to give an account of them.

For when a man knows not his own first principle, and when the conclusion and intermediate steps are also constructed out of he knows not what, how can he imagine that such a fabric of convention can ever become science?

-- Plato
Republic

Now, there are certain philosophers who, as we have intimated, themselves both affirm that it is possible that the same thing may and may not be, and that they really think so. [!]

This principle, however, do many investigators of nature employ. But we just now have assumed it as a thing impossible, in the case of an entity, that it should be and not be at the same time; and by means of this have we demonstrated that this is the most firm of all first principles.

-- Aristotle
The Metaphysics

It is always good to remind ourselves of the facts, to remind ourselves of what we actually know.

This method [*conceptual* revision], I believe, is one of the ways to make progress in philosophy. When confronted with an intractable question such as is presented by a clash of convincing default positions, don't accept the question lying down. Get up and go behind the question to see what assumptions lie behind the alternatives the question presents. In this case [dualism vs. materialism], we did not answer the question in terms of the alternatives presented to us but we *overcame* the question.

The fact that there are no universally accepted procedures for solving philosophical problems does not mean that anything goes, that you can say anything or that there are no standards. On the contrary, precisely the absence of things as laboratory methods to fall back on forces the philosopher to even greater degrees of clarity, rigor, and precision. In philosophy there is no substitute for a combination of original, imaginative sensibility, on the one hand, and sheer intelligent, logical rigor on the other. The rigor without sensibility is empty, the sensibility without rigor is a lot of hot air.

-- *Mind, Language, and Society*
John R. Searle

The model drawn as a hypothesis, as a tentative tracing, acquires once drawn the appearance of permanence, solidity, and reality --- so much so that even researchers who begin with an awareness of the hypothetical nature of their models often end by believing in their reality.

All disciplines tend to reify their abstractions, to mistake their main conceptual terms for concrete things. From long residence in our minds, our most familiar terms --- no matter how abstract they are --- are likely to be seen closer to home and hence more concrete than whatever we're supposed to be investigating "through" them.

-- *Words and Values*
Peggy Rosenthal

"For the want of a nail … the kingdom was lost."

Postmodern physics:

One could argue that physics is currently in the death throes of its postmodern phase with its antediluvian creation myth and a collection of just so stories concerning non-locality and reverse causality.

All jokes, hoaxes, and irony aside, we shan't go there.

At least 'the they' of modern physics have eliminated the completely unacceptable and 'occult' action at a distance from quantum field theory! They also replaced the improbable notion of electron spin with a collection of totally impossible ghostly supporting structures to shore things up.

For want of spin … all physics was lost !

In our new model spin is real. Angular momentum is the fundamental unit of interaction and the defining characteristic of photons and leptons. All physics is cast in terms of angular momentum and angular variables.

As a result, we express Newton's laws in terms of angular momentum in the center of mass of an N body system. This formulation is valid not only in any inertial reference frame, but in all accelerating reference frames as well. (See the next section.)

In our model, particle spin and planetary spin, etc. are instances of absolute motion that cannot be denied, or 'transformed' away. We fix our inertial reference frame with respect to the fixed background of space.

The results of quantum field theory and general relativity (will) both 'reduce' to our new model.

In addition, although our model was constructed with hindsight, one can derive the rules of quantum mechanics from our physical theory of particle propagation and interaction.

One cannot derive quantum mechanics from the Copenhagen interpretation.

We will begin this book with several pages of 'summary material', followed by the introduction.

beware *the 'they'* !

Interlude: More Epigraphs

On the Method of Mathematics:

The Euclidean method of demonstration has brought forth from its own womb its most striking parody and caricature in the famous controversy over the theory of *parallels*, and in the attempts, repeated every year, to prove the eleventh axiom. This axiom asserts, and that indeed through the indirect criterion of a third intersecting line, that two lines inclined to each other (for this is the precise meaning of “less than two right angles”), if produced far enough, must meet. Now this truth is supposed to be too complicated to pass as self-evident, and therefore needs a proof; but no such proof can be produced, just because there is nothing more immediate. . . .

In fact, it seems to me that the logical method is in this way reduced to an absurdity. . . .

But that axiom is a synthetic proposition *a priori*, and as such has the guarantee of pure, not empirical, perception; this perception is just as immediate and certain as the principle of contradiction itself, from which all proofs originally derive their certainty.

On Man’s Need for Metaphysics:

We also find *physics*, in the widest sense of the word, concerned with the explanation of phenomena in the world; but it lies already in the nature of the explanations themselves that they cannot be sufficient. *Physics* is unable to stand on its own feet, but needs a *metaphysics* on which to support itself, whatever fine airs it may assume toward the latter. For it explains phenomena by something still more unknown than are they, namely by laws of nature resting on forces of nature . . .

--- Arthur Schopenhauer
The World as Will and Representation

Newton's laws: (From "Revenge of the Sinister Universe")

In this section we present a synopsis of the 'universal' formulation, or expression, of Newton's laws, as proposed in our last book, "On Rotation".
[This formulation is good in *any* reference frame.]

Newton's first law:

If a particle does not experience a change in *angular momentum* relative to *any* arbitrarily chosen point in the universal reference frame, then the particle is considered to be free (i.e. there are no net forces acting on it).

For a two body system, if there is no change in the angular momentum of *either body* comprising the two body system relative to *any* arbitrary point (this 'excludes' the choice of points lying along the unit vector, **r**), then the particles are *not interacting*.

Newton's second law:

A particle that undergoes a change in angular momentum relative to our arbitrarily chosen point (i.e. the origin of our coordinate system) is said to experience a net torque, τ;

$$\tau = d\mathbf{L}/dt = \mathbf{r}\mathbf{x}\mathbf{F} \qquad \text{(a)}$$

where the force, **F**, is defined by

$$\mathbf{F} = d\mathbf{p}/dt = d(m\mathbf{v})/dt = m\, d\mathbf{v}/dt + \mathbf{v}\, dm/dt \qquad \text{(b)}$$

For two body 'central force' motion, the torques experienced by each individual body relative to our chosen reference point, are equal and opposite;

$$\mathbf{r}_1\mathbf{x}\mathbf{F}_1 = -\mathbf{r}_2\mathbf{x}\mathbf{F}_2 \quad ; \quad |\mathbf{F}| = |\mathbf{F}_1 - \mathbf{F}_2| \qquad \text{(c)}$$

Newton's third law:

During a two body interaction, the two bodies will undergo equal and opposite changes in their respective '*actions*'; i.e. they will have equal, and 'opposite', changes in kinetic energy.

$$\delta \int d\mathbf{L}/dt \bullet \omega dt = 0 \qquad (d)$$

where

$$\tau_{TOTAL} = \tau_{FORCE} + \tau_{SPIN} \qquad (e)$$

and

$$\tau_{SPIN} = \mathbf{I_1 x B_2} + \mathbf{I_2 x B_1} \qquad (f)$$

Newton's universal law of gravitation, expressed in the center of mass of a two body system, becomes; [note: equation (h) has been corrected relative to previous books.]

$$F/E_{TOT} = K^*(c/R)^2\mu - K^*(\mu v^2/R^2) - K^*(c^2 l^2/\mu R^3) \qquad (g)$$

$$K = (G/c^2) \qquad (h)$$

where the second term on the right hand side of equation (g) is the *coriolis* force; our answer to spacetime disturbances. [This extra term will describe the rotation of galaxies without the inclusion of 'dark matter'; and the perihelion of the planet Mercury.]
[For galaxy rotation, one *must* include the 'relativistic' mass of the individual bodies due to spin.]

Since our new coriolis force term goes as $1/R^2$, the classical and quantum mechanical conservation of the first three integrals of the motion, E, L, L_z, is still guaranteed.

Equation (d) represents the classical 'minimization of the action' expressed in our new angular variables. Relativistically, all interactions minimize the change in kinetic energy and equation (d) becomes

$$\delta \int (m(t) - m_0)\, dt = 0 \qquad (i)$$

In the final formulation, equation (g) will also have to include a spin term.
Particle [electron] in a box: (From "On wave particle duality ...")

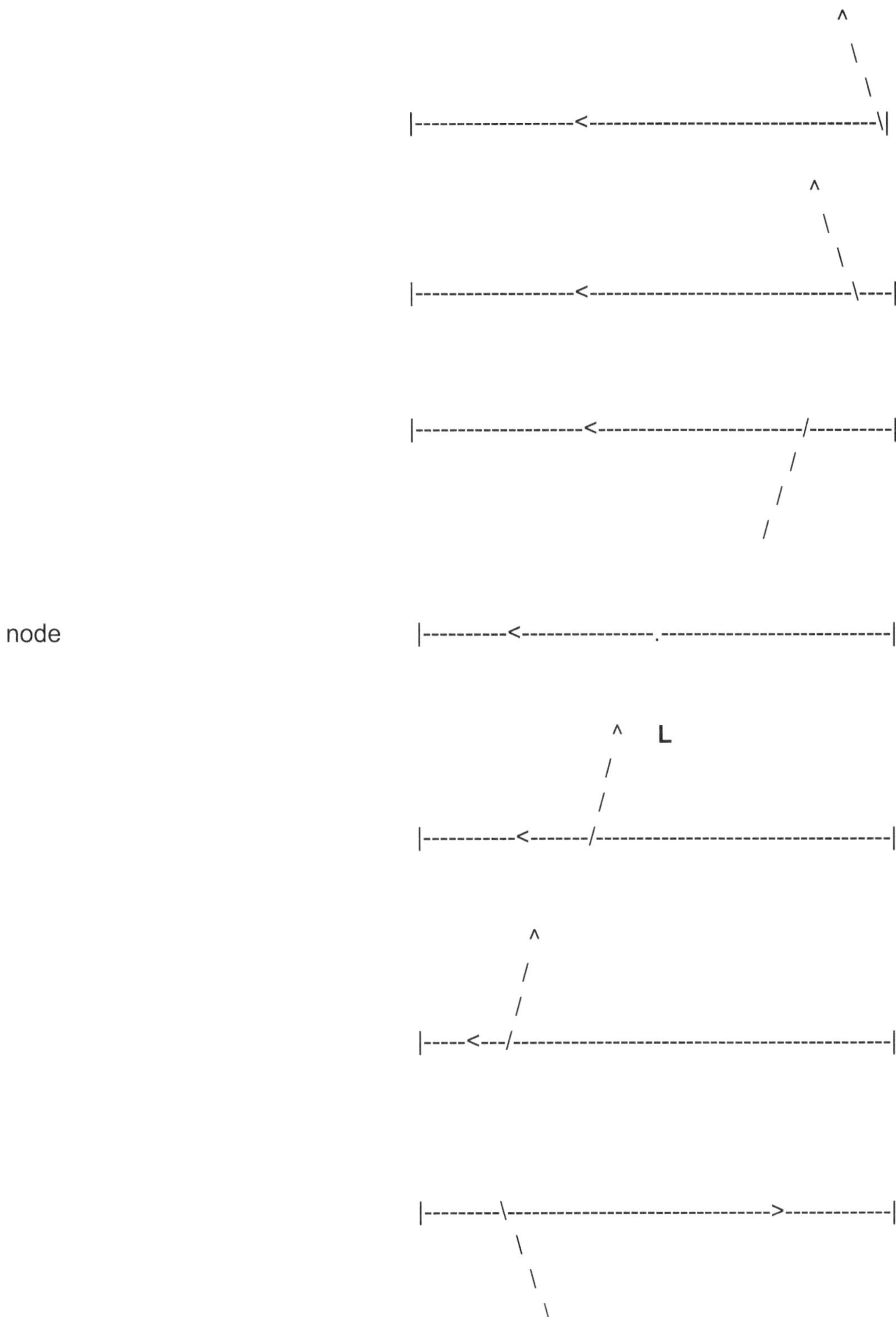

Figure A: At the ‘node’, the angular momentum vector **L** points up out of the page.

The neutrino: (From “On Rotation”)

We imagine the neutrino to be one quantum of action per unit volume of space, resulting in an available, or ‘functional’, angular momentum of $h^{bar}/2$.

The neutrino is a spinning point particle with three angular momentum vectors.

Only one component of the angular momentum may be projected along the direction of travel, and the total angular momentum ‘gimbles’ about this direction, as do the other two components. [The average of the other two components will obviously be zero!]

The gimbling of the neutrino [a spinor] is illustrated in Figure B.

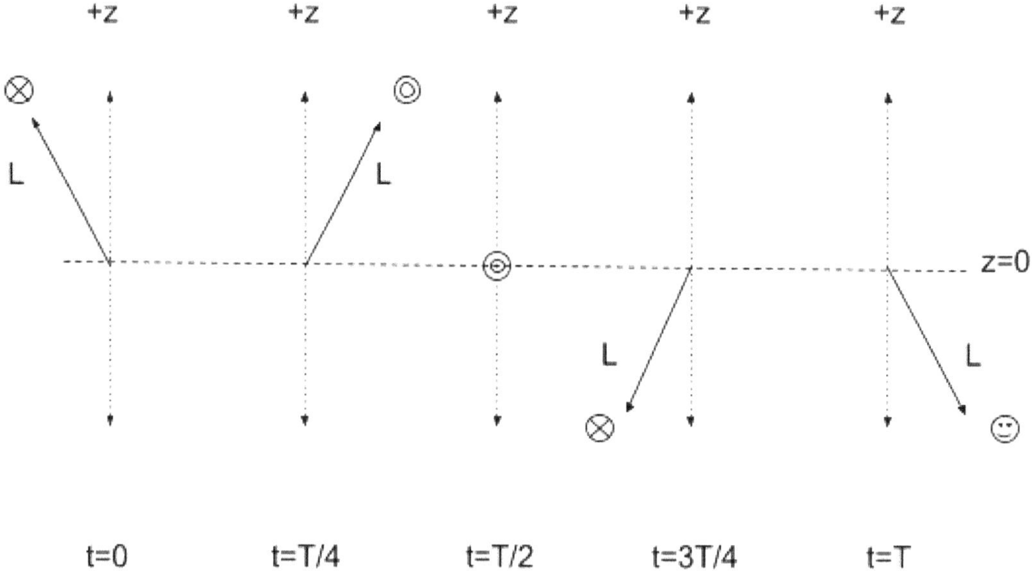

Figure B: At rest with a lepton traveling in the z-direction. The spin angular momentum vector ‘precesses’ about the direction of motion, tracing out a closed, three dimensional figure eight. The x symbol represents motion into the page. The dot symbol represents motion out of the page. At time T/2, we see the angular momentum is *perpendicular* to the direction of travel. (This is when the lepton engages in ‘virtual’ interactions.)

To represent an antilepton, simply swap the x symbols and the dot symbols.

The photon: (From "On Rotation")

The photon is one unit of angular momentum, *the fundamental unit of angular momentum*, Planck's quantum of action; h. The angular momentum of the photon is always h, and the speed is always c. So, how does the photon gain and lose energy and linear momentum during an interaction?

In our model, the projection of the photon angular momentum vector, along the direction of travel, varies sinusoidally at the frequency that defines the energy and momentum of the photon according to the usual relations;

$$E = h\upsilon$$
$$p = h\upsilon/c \qquad (j)$$
$$m = h\upsilon/c^2$$

Let us say the photons 'gyres' as illustrated in Figure C.

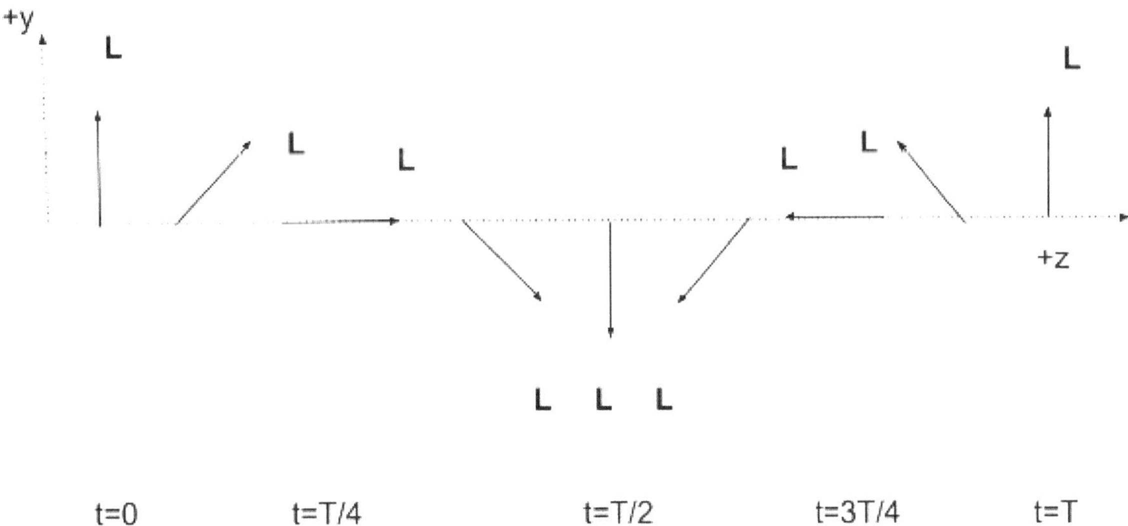

Figure C: The photon rolls, or 'gyres', along the direction of travel projecting the spin angular momentum vector sinusoidally along the direction of propagation while maintaining a constant polarization and angular momentum; L = h. We could also say that the photon '**spirals**'. At t=0, the plane of the photon spin is *coplanar* with the z-axis.

A universal metaphysics ?: (From “On Epistemology and Ontology”)

The fundamental unit of matter is the neutrino.
The neutrino is a gravitational-point-mass-charge;
or, a point, gravitational, magnetic moment, μ

$$\mu = h^{bar}/2c \, (1 + \tfrac{1}{2} v^2/c^2 + \ldots .) \qquad \text{(k)}$$

The elemental unit of matter is the dipole! This may spark thoughts of ‘dialectic’ in our philosophical readers; however, the magnetic dipole is represented, mathematically and schematically, as closed loops of ‘magnetic field lines’ ! This picture of the neutrino is illustrated in Figure D.

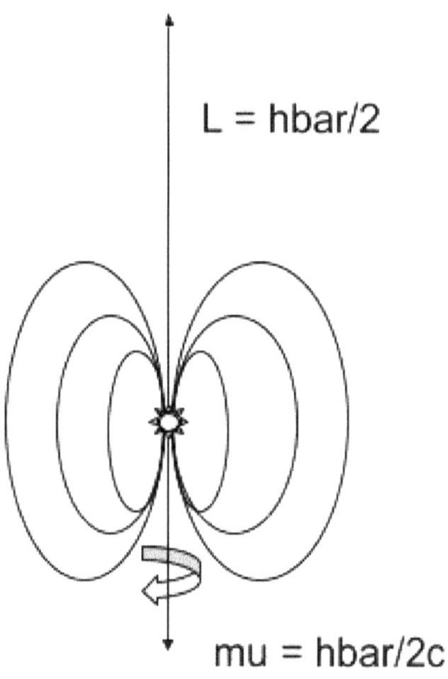

Figure D: The angular momentum, magnetic moment, and magnetic field lines of a point, gravitational, mass dipole; aka the neutrino!

The neutrino is spinning to the left, with the canonical definitions for the directions of L and μ given by the standard model.

Thus, there are no magnetic monopoles !

The electromagnetic charge: (From “Revenge”)

Interactions conserve spin, mass, and charge. In the universal model, we have generalized the concepts of electric charge and mass, to include all conserved quantities.

We define the electromagnetic charge to be

Q_{EM} = e h^{bar}/ 2 c **s** (l)

and the mass charge to be

Q_{MASS} = m h^{bar}/ 2 c (m)

These quantities must be conserved at every ‘vertex’.

The nature of these new charges are summarized in Table A.

Force	Coupling constant	Conserved current	Rotation basis	Conserved charge
Electricity	e/m_e	Mass isospin	e,mu,tau	e*hbar/2*c **s**
Gravity	m_e/e	Charge isospin	e, υ_e	m*hbar/2*c

TABLE A: Table of coupling constants, conserved currents, and charges: Note: $m_e/e = m_\upsilon$

The leptonic table: (From “Revenge”)

LEPTONS			ANTI-LEPTONS	
electron	electron neutrino	PARITY ⇔	electron antineutrino	positron
⇐	CHARGE	MASS ⇅	CHARGE	⇒
muon	muon neutrino	PARITY ⇔	muon antineutrino	anti-muon
⇐	CHARGE	MASS ⇅	CHARGE	⇒
tau	tau neutrino	PARITY ⇔	tau antineutrino	anti-tau
⇔	mass isospin	charge isospin ⇕	mass isospin	⇔

TABLE B: The leptons and their interrelations; or the kleptogenesis of the leptoquarks.

Any lepton can be ‘generated’ from any other by the appropriate applications of the parity operator, the mass isospin operator, and our newly proposed ‘charge isospin’ operator.

Using various combinations of the step up and step down operators of SU(2) [charge isospin] and SU(3) [mass isospin], plus the parity operator, we can write any quantum mechanical interaction current in terms of the ‘fundamental’ neutrino neutral current.

Units and dimensions:

It is time to take the plunge and redefine the electric charge in terms of mass. This is the last necessary step to put gravity and electromagnetism on an equal footing. In our model, the fundamental coupling charge for all interaction is the particle mass. The charge to rest mass ratio of the electron, e/m_e, is now a coupling constant that acts to distinguish the electron from the neutrino. The new 'electric charge' is $m(e/m_e)$. We can see that the constant e must now have units of mass to make the new units work out correctly.

We believe that the phenomenon of electromagnetic induction is responsible for the increase in electron mass relative to the neutrino and endows it with an electric charge and thus an additional electric field. It is the motion of mass, rather than electric charge, that induces magnetic forces in our new model leading to the hypothesized magnetic moment of the neutrino. Obviously, this will involve the redefinition of several units, making them all Coulomb free. Mass currents and electrical currents will then be on equal footing, and will interact with one another, but with very different coupling strengths, of course.

We have yet to work out the details of what these new units might be called and how they will affect other units and their designations.

Details ! See "On Rotation" for a few more thoughts and details on this subject.

Errata:

We have conjectured (1) that the value or magnitude of e is the ratio of the mass of the electron to the mass of the neutrino

$$e = m_e/m_\upsilon \qquad \text{(o)}$$

We have decided this is (obviously!) wrong as it does not work out mathematically or jibe with our new redefinition of the units of e as described above. Why I have held onto this idea so long although I could see there was something wrong with it is a mystery to me.

sigh

Introduction:

The universe is governed by one force; gravity or electromagnetism, call it what you will.

There are two fundamental matter particles; the neutrino and the electron.

The difference between the strengths of gravity and electromagnetism, and the difference between the mass of the neutrino and the mass of the electron, are, both, due to the same fundamental phenomenon; electromagnetic induction.

Electromagnetic induction is the 'symmetry breaking mechanism' of the universal model.

Electromagnetic induction is also the most 'difficult' concept in our model, and is totally derived from the classical idea. We take it to be an empirical discovery, and thus 'irreducible' theoretically.

The earth is a terrible place to do physics. It is round and it spins. Locally, it is flat and full of friction; air resistance, viscous fluid flow, sticky surfaces, rough terrain, etc.

We have found during our investigations that we must agree with Aristotle, and many other ancient Greeks: the natural state of a particle is to be 'at rest', and all particles that are in motion *tend toward* a state of rest even though nature is nothing more than particles in motion, and/or constant change and becoming! More on this later !

In our model, the total force between any two objects (electrons, planets, etc.), normalized by the total energy of the system, and cast in planar coordinates is

$$\mathbf{F}/E_{TOT} = K^*(c/R)^2\mu - K^*(\mu v^2/R^2) - K^*(l^2/\mu R^3) \qquad (1)$$

where μ is the reduced (relativistic) mass, and

$$K == (G/c^2 - (e/m_e)^2(\mu_0 /4\pi)) \qquad (2)$$

Equation (1) is essentially all of physics in a nutshell ! :)

That is the premise of the universal model.

Symmetry:

The symmetries of the universal model are; up, down, left and right --- essentially, the conservations of parity and helicity. The relationship between these quantities is conserved under rotations. The invariance of any system under rotations insures the conservation of angular momentum and kinetic energy; the only two *truly fundamental physical quantities.*

The invariance of systems under linear transformations insure, or demonstrate, the conservation of linear momentum *and* the conservation of energy. This is so because in our model the position and momentum coordinates are variables, while *time* is merely a parameter. Time is essentially over *constrained* and can not vary independently relative to the space coordinates. In addition, the concept of potential energy has been 'discredited' in a technical sense, although perhaps not operationally. Hence there is no need for a time translational invariance to demonstrate the conservation of energy. ALL energy is kinetic energy.

Due to their nature as *spinors*, fermions that spin to the left, spin to the left whether or not they are pointing up or down. They spin to the left whether or not the spin is projected along the direction of propagation or against it. Hence, one cannot change the helicity of a particle by making a Lorentz boost as is commonly believed.

Fermions that spin to the left are *matter,* and have a parity of +1. Fermions that spin to the right are *antimatter,* and have a parity of -1.

Protons *spin to the right.* Protons have a parity -1. One may call them antimatter, if it really matters! All this has been explained in our previous books, so we won't go into the details of this determination here.

The essentially random assignment of a parity of +1 to the proton in the standard model has led to all kinds of excitement and confusion concerning parity and CP violation.

A tempest in a teapot. Much ado about nothing !

As for *gauge symmetry,* it is just not a thing. As we have shown, mass, or mass-energy, is the fundamental coupling charge. The rest mass of a particle corresponds to the 'global charge' of the original electromagnetic gauge theory, and the relativistic mass is the 'local charge.' The universal wave function is then

$$\psi = \exp(i(p \bullet x - (m-m_0)c^2t)/h^{bar}) \qquad (3)$$

and, the Dirac equation is

$$ih^{bar}\partial\psi/\partial t = -ich^{bar}\alpha \bullet \nabla\psi \dashrightarrow mc^2\psi = c\alpha \bullet p\, \psi + m_e c^2\psi \qquad (4)$$

Cause and effect:

The *inter*action between any two objects (or collection of objects) is due to the equal and opposite forces objects exert on each other as indicated in equation (1). All *inter*action is action and reaction. As Kant has taught us, the idea of causality is an *a priori*, or intuitive, conception imposed on events by consciousness or the mind. In more modern terms, we would say this way of perceiving the world is hardwired in the animal brain due to hundreds of thousands, if not millions, of years of evolution. The idea of causality arises from the temporal nature of these interactions. Time flows forward by definition. Don't let anyone tell you otherwise!

The confusion about cause and effect and the 'direction' of time began early in the history of physics and predates both quantum theory and relativity theory. One wants to describe the motion of a projectile in the earth's gravitational field. We know the projectile is going to travel *forward* in both space and time so we extract the appropriate physical parameters from our essentially metaphysical notions of space and time and construct the equations of motion.

Then, some Clever Trevor notes that one can put a minus sign in front of every t variable leaving the equations of motion 'unchanged' and thus concludes it is some great mystery why particles do not travel backwards in time. *There is no mystery.* The equations in question were specifically designed to describe forward motion. The equations are simply a tool designed by people to describe specific observations and physical situations. They have no meaning or inherent value or significance themselves. The idea that time flows forward was a preconception and condition in the construction of the equations. One can easily fix the mystery by defining time to be 'positive-definite.'

The real mystery is why *anyone* would find this interesting, confusing, exciting, or indicative of some deep conundrum concerning the physics of time.

The equal and opposite forces of action and reaction between two particles or 'objects' are due to the constant and continuous exchange of energy and momentum by way of the virtual photon connecting the two. The frequency of the virtual photon, and thus energy exchanged, is proportional to and varies with the distance between the objects. This variation in frequency, and thus the transmission of the varying force between the two, is 'instantaneous' and requires the specification of a common measure of time. This model is in stark contrast to modern field theories which assume that the propagation of internal forces occur at a finite speed. One object emits a virtual particle and the other absorbs it. But which is the source and which is the sink? The more massive object is always the source, of course! Modern field theory fails to recognize the fundamental difference between real and virtual photons. Only real photons are emitted and absorbed, and travel through space at a finite speed.

Time and space:

Oh my gosh! It is going to take teams of psychologists, sociologists, historians, philosophers, and, yes, even physicists, many, many years to unravel and understand how people have arrived at their current incoherent and childish conceptions concerning time and space.

We certainly do not have the time, patience, or inclination to tackle the issue here. Physicists should stick to physics and leave matters on ontology, epistemology, and mathematics (see the disaster that is gauge theory!) to the respective professionals. In addition, mathematicians should not try their hand at physics. Of course, the pursuit of physics touches upon all these pursuits, particularly mathematics, but one must always start with a physical model. Do the physics, then the math, and then the philosophy. The true scandal is the manner in which philosophers have capitulated to the physicists and try to lend support and justification to all their pathetic ideas.

Philosophy is no longer the handmaiden of physics. Those days are long gone and ended with the Age of Reason.

As for time and space, one cannot speak of such things in terms of physics without respect to a given problem and choice of reference frame.

Whether one is studying the interaction between two fermions, planets, or galaxies, one must determine the center of mass, establish a fixed reference frame with respect to this center and then employ the Newtonian conceptions of time and space.

Take the interaction of two galaxies, which is bound to raise the most immediate objections.

These two objects are moving with respect to each other and due to the influence of the one on the other by the equal and opposite forces of action and reaction. It does not matter how far apart they might be. Put your clock at the center of mass and measure!

We have touched upon why current ideas of time and space are now such a clusterfuck in several previous books, but not in a rigorous and exhaustive way. Currently, we have bigger battles to fight and fish to fry! In addition, it is hard to know quite where to start to correct misconceptions that are so obviously wrong. Unfortunately, people currently seem quite proud of the fact that physics is crazy and mind boggling. Mind boggling! Oh my gosh!

Coming soon: "On Time and Space" and "On TIme and Temperature".

Interlude: More Epigraphs !

Thus energy, like the atom, is more and more divested of all sensuous meaning with the advance of knowledge. This development appears most clearly in the concept of potential energy, which even in its general name points to a peculiar logical problem. As Heinrich Hertz has emphasized, there is a peculiar difficulty in the assumption, that the alleged substantial energy should exist in such diverse forms of existence as the kinetic and the potential form. Potential energy, as it is ordinarily conceived, contradicts every definition that ascribes to it the properties of a substance; for the quantity of a substance must necessarily be a positive magnitude, while the totality of potential energy in a system is under some circumstances to be expressed by a negative value. Such a relation can, in fact, according to Gauss' theory of negatives, only be explained where that which is counted has an opposite, *i.e*, "where not substances (objects conceivable in themselves) but relations between two objects are counted."

⇔

Thus the passive *fixity*, established by science at certain points, is an element in its own activity. In fact, it is justified and unavoidable, that science should condense a wealth of empirical relations into a single expression, into the assumption of a particular thing-like "bearer." The critical self-characterization of thought, however, must analyze this product once more into its particular factors, although it conceives this product as *necessary* for certain purposes of knowledge. This is done because critical thought is not directed forwards on the gaining of new objective experiences, but backwards on the origin and foundations of knowledge. The two tendencies of thought here referred to can never be directly united; the conditions of scientific *production* are different from those of critical *reflection*. We cannot use functions in the construction of empirical reality and at the same time consider and describe them.

⇔

The goal of theoretical physics is and remains the universal laws of process.

--- Ernst Cassirer
Substance and Function

Math and physics:

The reason mathematics describes physics so well is because the two were devised in tandem, *and* because physical processes are mechanistic and deterministic.

Now, when physics tries to import independent mathematical ideas such as group theory or non-Euclidean geometry, the result is an unmitigated disaster. There are no 'internal symmetries' dictating the nature of physical processes, and for all intents and purposes space is Euclidean and independent of time.

Even well motivated and useful mathematical abstractions can lead to disaster when people begin to take them too seriously. The perfect example is the concept of the field and its antecedent concept; potential energy. As we have shown, potential energy is a useful placeholder for many mathematical operations but is not a real thing that inheres in particles, or a system of particles, or in fields.

The concepts of the field and potential energy originate in Classical physics, of course. Physicists have 'forgotten' that fields are a mathematical abstraction, or calculational convenience, *derived from the equal and opposite forces observed between two isolated interacting objects*. This abstraction is useful when one wants to study special, idealized situations. However, the idea of taking the limit of the force per unit charge as the value of a 'test charge' goes to zero is mathematically problematic, even in the Classical formulation. I don't think this problem has gone away.

Let's take classical electromagnetic waves as an example. A plane electromagnetic wave is not a series of sinusoidally oscillating electric and magnetic fields of fixed frequency. In our formulation, such a wave is a series of wave fronts consisting of photons. The photons themselves oscillate at the same frequency that determines their rate of emission from a given source.

The description or reduction of these photons in terms of electromagnetic waves is analogous to the characterization of the microscopic properties of an ideal gas in terms of the macroscopic variables of volume, temperature and pressure.

Now the universe is rife and lousy with fields of every description.

New problem? New field!

The main sin and major strike against General Relativity is that it is a field theory.
In addition, it has an 'infinite' number of predictions, many unphysical.

Falsification defined.

Physics and metaphysics:

'Modern' physics is replete with metaphysical assumptions whether the current devotees, or practitioners (if we are being gracious), care to admit it or not. In truth, one cannot call such a hodge podge of kludges and ad hoc assumptions metaphysics proper, as they reflect no underlying unifying or consistent philosophy.

Disembodied fields, collapsing wave functions, the vacuum, broken internal symmetries, particles traveling backward in time, etc., are not only silly ideas, but they can never be observed or empirically verified. The main argument for their reality is they make the math work out right. Given the number of tunable knobs in the standard model, plus renormalization, it is no wonder that people can make so many things work out to their satisfaction.

Currently, our new model is based on one simple *metaphysic*; particles 'want' to be at rest or achieve the lowest possible kinetic energy given the local environment. (It is still hard to express such ideas without using anthropological terms!)

All particles are connected by virtual photons; one virtual photon per particle pair. These photons are unbreachable and unbreakable. Each particle in the universe sees every other particle as a possible sink into which to dump their excess energy and momentum. This ongoing contest of mutual exchange causes particles to accelerate and gravitate towards or away from one another depending on their individual charges and spins.

This minimization process is expressed mathematically in equation (i) which we will reproduce here

$$\delta \int (m(t) - m_0)\, dt = 0 \qquad (5)$$

Photons and leptons:

In our model photons have mass as indicated in equation (j), which we reproduce here

$$m = h\upsilon/c^2 = h^{bar}\omega/c^2 \qquad (6)$$

Of course, the photon also has a momentum which we will write as

$$p = h^{bar}\omega/c \qquad (7)$$

The work done on a photon can be expressed as follows (see "On Rotation")

$$\int F\,dt = \int (dp/dt)\,dt = \int (h^{bar}/c)(d\omega/dt)\,dt \qquad (8)$$

In our model the universe is not expanding, and the shift in frequency of light from a distant galaxy should be derived from equation (8) where the force is due to the mass of the emitting galaxy, formulated from Newton's universal law in the usual way. There will be an additional complication due to the fact that the separation between the galaxy and the photon will be time dependent; R = R(t). We shan't explore this further right now!

As demonstrated/hypothesized in "On Rotation" the **photon** also has a **magnetic moment!**

$$\mathbf{m}_\gamma = h/c\,\mathbf{i} \quad ; \quad \mathbf{i} = \mathbf{v}/|\mathbf{v}| \quad ; \quad v = c \qquad (9)$$

[equations (8) and (9) have been corrected] and, finally, the photon has a moment of inertia

$$I_\gamma = (h^{bar}\omega/c^2)(\lambda^2/4\pi^2) \qquad (10)$$

In our model the photon has an 'effective radius' equivalent to its wavelength, λ,as imagined in Figure C. We believe that this is how the photon is able to go through two slits of a similar separation "at the same time."

The **neutrino** also has a **magnetic moment**. This has been derived by an analogy with the magnetic moment of the electron (over the course of our investigations) and is given in equation (k). This magnetic moment is solely due to and is implicitly dependent on the spinning mass.

Accordingly, we find that the **electron magnetic moment** must have an additional magnetic moment term due to the mass contribution

$$\mu_e = (e/m_e + 1)(h^{bar}/2c)(1 + \tfrac{1}{2} v^2/c^2 + \tfrac{3}{8} v^4/c^4 + \ldots) \qquad (11)$$

Note that the magnetic moment is velocity dependent. If we neglect the contribution from the mass term for now then equation (11) reduces to

$$\mu_e = (e\, h^{bar}/2m_e c)(1 + \tfrac{1}{2} v^2/c^2 + \tfrac{3}{8} v^4/c^4 + \ldots) \qquad (12)$$

and we can see in our model the 'anomalous' correction of the standard model is cast in terms of an expansion in $(v/c)^2$ rather than α. Remembering that one of the definitions of $\alpha$is in terms of the velocity of the electron in the ground state of the hydrogen atom, it seems these two formulations could be reconciled. We shall not pursue this further right now. We also note that in our model the running of $\alpha$is also expressed in an expansion of $(v/c)^2$.

$$\alpha = \alpha_0(1 + (v/c)^2 + (v/c)^4 + \ldots) \qquad (13)$$

$$\alpha_0 = e^2/4\pi\varepsilon h^{bar} c \qquad (14)$$

Whether this is "covariant" or whether it matters we shall also not consider right now.

We remind the reader that in our model the factor of e/m_e is dimensionless so the magnetic moment of the electron (and the photon and the neutrino) has units of length•mass and can be defined as "the first moment of mass." We have defined the 'radius' of the electron at rest to be equivalent to the Compton wavelength, so, we can write for the first moment of mass

$$\lambda_e m_e = (h/m_e c)m_e = h/c \qquad (15)$$

In this simple hand waving argument, at least the units come out right!

Finally, we note that in our model the second moment of mass, or the moment of inertia, is

$$I = m\lambda^2/(2\pi)^2 \qquad (16)$$

This formula applies for the electron and the neutrino and the photon.

The weak and strong forces:

While we are on the subject of the running of the magnetic moment and the electromagnetic coupling constant it seems a good time to (re)introduce the reader to our derivation of the coupling constants for the other three forces.

The gravitational coupling constant is

$$\alpha_G = (m_e^2 G)/(h^{bar} c)\ (1 + (v/c)^2 + (v/c)^4 + \dots) \qquad (17)$$

The weak coupling constant is

$$\alpha_W = (m_\nu^2 G)/(h^{bar} c)\ (1 + (v/c)^2 + (v/c)^4 + \dots) \qquad (18)$$

And finally, the strong coupling constant (14) is

$$\alpha_S = (G/4\pi\varepsilon)^{1/2}\ (2 m_e e/h^{bar} c)\ (1 + \tfrac{1}{2} v^2/c^2 + \tfrac{3}{8} v^4/c^4 + \dots) \qquad (19)$$

In the universal model, all the ‘constants’ run, because *the fundamental coupling charge* of a particle *is the relativistic mass-energy* of the particle. There is no weak charge or color charge in our model and the weak and strong forces are simply the manifestation of gravity and electromagnetism at subatomic scales and relativistic velocities. Obviously, the weak force is simply the gravitational force and is weak due to the mass of the neutrino!

There is no need for renormalization.

In our model the recently discovered Higgs boson is the gravitationally bound state of a neutrino and an antineutrino. Each neutrino can radiate a (‘virtual’) photon, each of which produces a lepton antilepton pair, or the neutrino and antineutrino can ‘annihilate’ to produce a virtual photon then producing a lepton antilepton pair leading to the following decay modes;

$$H \rightarrow \nu\nu^{bar} + l^+l^- + l^+l^- \ ;\ l = e, \mu, \tau \qquad (20)$$

$$H \rightarrow l^+l^- \ ;\ l = e, \mu, \tau \qquad (21)$$

In equation (21) the virtual photon can also produce a photon anti-photon pair!

With the proper choice of radius, using the known mass of the neutrino, and applying the uncertainty principle, it should be easy to estimate the mass of our Higgs. We leave this as an exercise for the reader!

The propagator:

In our model the mass dependence, and short range, of the 'weak' force is due to the tiny mass of the neutrino in the coupling parameter (plus the factor of G) as shown in equation (18) and is *not* due to a massive exchange boson as in the standard model. The only exchange boson is the virtual photon. Technically, the virtual photon is not exchanged, but represents a constant tethering of two particles, allowing for a continuous exchange of energy and momentum consisting of 'equal and opposite action and reaction'.

The idea of a constant and continuous exchange of energy and momentum between two interacting particles is best illustrated in the construction of the propagator between two 'spinless' leptons in "Quarks and Leptons" (Halzen and Martin) where it is simply explained that the *current of either particle* can be chosen as providing an electromagnetic field influencing the other particle and *the resulting photon propagator between the two is the same*. We take this most sensible and excellent presentation from pages 84-87 and refer the reader there for all the details. Their derivation is in terms of the 'lowest order' approximation using field theory, although we shall adapt it to motivate our field free model.

In our reformulation, A^{μ} represents the 'wave function of the virtual photon' and not the electromagnetic potential. Hence, we needn't disregard the higher order contribution from the e^2A^2 term because it no longer is relevant. The transition amplitude is

$$T_{fi} = -i \int j_{\mu}^{1} A^{\mu} d^4x \quad ; \quad A^{\mu} = -(1/q^2)\, j^{\mu}_{2}$$

and the invariant amplitude is

$$M \sim e^2(-i\, g_{\mu\nu}/q^2) \quad ; \quad \text{where } -i\, g_{\mu\nu}/q^2 \text{ is the photon propagator.}$$

This is all very rough of course and we may have not needed to include the invariant amplitude to illustrate the differences between the standard approach and our new model, but it conveniently introduces the factor of e^2 into the discussion.

Now, besides the fact that our model is 'linear' in A^{μ}, the other important point to emphasize is in our model *the electric charge is mass dependent* and we must replace e with $m(e/m_e)$, where m is the relativistic mass. Hence, higher order contributions to the matrix element will not be due to or derived from a perturbative expansion in α *per se*, but will arise from the increase in the masses of the particles in high energy interactions. See equation (13).

The Dirac equation:

As we mentioned in the section on Symmetry, we have *successfully removed* the rest mass term from the traditional Dirac equation and placed it in the wave function instead. We believe this has several important consequences. Firstly, it should make the construction of the Dirac Lagrangian easier and more natural. Secondly, it may remove peoples' belief that having removed the rest mass term from the standard model Dirac Lagrangian 'by hand', that there is a need to reintroduce or reinstate the particle mass somehow by recourse to the Higgs field.

We're not sure why people believe crazy things, to be honest !

So the universal Dirac equation is

$$H \psi = c\alpha \bullet p \psi \qquad (22)$$

$$i\hbar \partial\psi/\partial t = -ic\hbar\alpha \bullet \nabla\psi \qquad (23)$$

and is satisfied by, or solved with, the universal wave function

$$\psi = \exp(i(p\bullet x - (m-m_0)c^2t)/\hbar) \qquad (24)$$

$$mc^2\psi = c\, \alpha \bullet p\, \psi + m_e c^2\psi \qquad (25)$$

The free particle, spin up, solution of the Dirac equation for an electron (positive helicity) is

$$\psi = \begin{vmatrix} 1 \\ 0 \\ \sigma \bullet \mathbf{p}/mc \; 1 \\ 0 \end{vmatrix} \exp(i(p\bullet x - (m-m_0)c^2t)/\hbar) \qquad (26)$$

$$\sigma \bullet \mathbf{p}/mc = \begin{vmatrix} p_z & p_x - ip_y \\ p_x + ip_y & -p_z \end{vmatrix} \qquad (27)$$

$$\rightarrow \text{the determinant} = -(p_z^2 + p_y^2 + p_x^2) \rightarrow -(p_z^2 + p_y^2 + p_x^2)/mc = -1 \qquad (28)$$

$$\rightarrow \text{the square} = (p_z^2 + p_y^2 + p_x^2)\, \mathbf{1} \;\; ; \mathbf{1} = \text{2x2 unit matrix} \qquad (29)$$

The matrix $\sigma \bullet \mathbf{p}/mc$ *is the factor* that turns our ordinary column vector *into a spinor,* which is technically classified as an Euclidean tensor. So the universal model involves scalars, vectors, and tensors: all Euclidean.

A spinor is a vector with zero 'length' which means that two of the three components must be imaginary. In our example, only p_z is real; the direction of propagation!

Since the spinor precesses about the direction of propagation, the average values of p_x and p_y are zero, but this does not mean they are undetermined or do not have distinct values and directions at any given time.

The form of equation (21) turns up again and again; i.e. rotations, angular momentum … however, equation (21) is not the result of a rotation or 'cross product', as in the other examples.

We hope to explore the role and significance of complex numbers in modern physics more in our next book.

The Dirac equation with the universal interaction term added is

$$i\hbar \partial\psi/\partial t = -ic\hbar\alpha \bullet \nabla\psi + mc^2 A^\mu \qquad (30)$$

where A^μ represents the wave function of the virtual photon mediating the interaction with a second particle. It does not represent a potential field. For a single particle interacting with several other particles, A^μ would be represented by a Fourier sum or integral over all the relevant virtual photon frequencies.

no more fields

[This section has been cut and pasted from "Toward a Metaphysics of Mass and Motion"]

[Sorry for the Clip Show!]

Postmortem physics:

Let us be brief.

Fields are not real. The quantum mechanical vacuum is not real. Potentials are not real. Gauge symmetry is not real. Potential energy is not real. The Higgs boson is not real.

Wave functions are not real.

None of these things are real.

Mass is real. Spin is real. Kinetic energy is real. Acceleration is real.

Photons are real. Electrons are real. Neutrinos are real. They are all mass in motion.

There is only mass in motion.

We can't even, as they say.

: /

Conclusion:

This book did not turn out quite as expected, but then they never really do. We haven't calculated much of anything nor introduced too many new ideas. The Postmortem was not the vivid vivisection or blistering rebuttal of the standard model that we had envisioned. I become weary just thinking about it. I weary of wondering why people think it is so wonderful. I say, let the books speak for themselves.

I say, let people think for themselves. It is possible! I must believe it.

So, let us consider this book an intermezzo.

In our next offering(s) we expect to take a closer look at spin, the Born rule, the Born approximation, the uncertainty principle, the Dirac equation, and the Hydrogen atom, etc. I don't expect we will still actually calculate much of anything, but I do expect we will finally finish laying a firm foundation for our new 'universal model' and provide the tools for others to do useful work.

Books by Greg Feild:

1. “A quantum mechanical theory of gravitational interactions”
   CreateSpace Independent Publishing, 8/29/2016

2. “Observations on the quantum mechanical nature of gravity”
   CreateSpace Independent Publishing, 10/8/2016

3. “On gravitation and electric charge”
   CreateSpace Independent Publishing, 10/29/2016

4. “On spin, mass, and charge”
   CreateSpace Independent Publishing, 11/29/2016

5. “On angular momentum, acceleration, and absolute motion”
   CreateSpace Independent Publishing, 1/1/2017

6. ”The Sinister Universe”
   CreateSpace Independent Publishing, 3/1/2017

7. ”On Parity and Isospin”
   CreateSpace Independent Publishing, 4/11/2017

8. “Reflections on the Sinister Universe”
   CreateSpace Independent Publishing, 5/12/2017

9. “On Current Physics”
   CreateSpace Independent Publishing, 6/11/2017

10. “A Critical Examination of Classical and Quantum Mechanical Waves”
    CreateSpace Independent Publishing, 6/18/2017

11. “On wave particle duality and the quantum of action”
    CreateSpace Independent Publishing, 7/6/2017

12. “On matter, mass, and motion”
    CreateSpace Independent Publishing, 9/14/2017

13. “On action and reaction”
    CreateSpace Independent Publishing, 9/24/2017

14. “A quantum mechanical theory of everything”
    CreateSpace Independent Publishing, 11/5/2017

15. “On Interaction”
CreateSpace Independent Publishing, 4/21/2018

16. “On Rotation”
CreateSpace Independent Publishing 8/19/2018

17. “Revenge of the Sinister Universe: The Reality of Everything’
CreateSpace Independent Publishing, 9/4/2018

18. “On Math, Physics, and Metaphysics”
CreateSpace Independent Publishing, 10/1/2018

19. “On Quantum Mechanics”
CreateSpace Independent Publishing, 10/15/2018

20. “On Epistemology and Ontology”
CreateSpace Independent Publishing, 10/21/2018

21. ”Toward a Metaphysics of Mass and Motion”
CreateSpace Publishing, 10/29/2018

Compilations:

A. “The Universal Model of Our Sinister Universe: The First Ten Books”
CreateSpace Independent Publishing, 7/2/2017

B. “The Canons of the Sinister Universe:
The Last Four Books on the Universal Model of Our World”
CreateSpace Independent Publishing, 11/5/2017

C. “The Return of the Sinister Universe: The Immaculate Collection”
CreateSpace Independent Publishing, 9/4/2018

D. “The Battle for the Sinister Universe: The Heuristics”
CreateSpace Independent Publishing, 10/29/2018

Resources:

Quantum Field Theory
Claude Itzykson, Jean-Bernard Zuber

Atomic and Quantum Physics
H. Haken, H.C. Wolf

Modern Elementary Particle Physics
Gordon Kane

Classical Dynamics of Particles and Systems
Jerry B. Marion

Foundations of Electromagnetic Theory
John R. Reitz, Frederick J. Milford, Robert W. Christy

Quantum Physics
Rolf G. Winter

Gauge Theories in Particle Physics
I. J. R. Aitchison and A. J. G. Hey

Quarks and Leptons: An Introductory Course in Modern Particle Physics
Francis Halzen, Alan D. Martin

Quantum Field Theory
F. Mandl, G. Shaw

Theoretical Mechanics of Particles and Continua
Alexander L. Fetter, John Dirk Walecka

The Theory of Spinors
Elie Cartan

Elementary Modern Physics
Richard T. Weidner, Robert L. Sells

Quantum Mechanics
Claude Cohen-Tannoudji, Bernard Diu, Franck Laloe

The misanthropic principle:

everyone is the worst !

News for parrots:

no parrots were involved.

↶↷

Mechanism and Energetics

Greg Feild

October 1, 2021

About the author

I earned a Ph.D in experimental high energy physics from the Pennsylvania State University working on HERA at DESY in Hamburg, Germany studying photoproduction and deep inelastic scattering in electron-proton collisions.

I did my postdoctoral studies with Yale University working at Fermilab on the CDF experiment at the Tevatron. My primary research interest was particle hadronization in quarkonium production in proton-antiproton collisions.

Abstract

Our contention is that all particles can be completely characterized in terms of angular momentum. The mass-energy of any and all particles is nothing more than, and can be completely described as, rotational kinetic energy. Particle interactions involve the exchange of rotational kinetic energy and angular momentum in the form of the photon. There are no other mechanisms necessary to describe the nature of our world. There are no other particles besides the photon, the three leptons, and their neutrinos.

Energetics

I first came across the term Energetics or Energism while reading “Substance and Function” by Ernst Cassirer, an engaging book about the Philosophy of Science written around the turn of the last century. In the book, he presents energism and mechanism as diametrically opposed approaches to the problems of physics. It seems Cassirer comes fully down on the side of energism although his contention is energy is a quantity that exhibits many forms that can not be modelled but only mathematically maintained, manipulated and measured. In more familiar terms, he advocates for the Hamiltonian approach over the Newtonian approach to kinematics.

Our model is a compromise or a marriage of these two approaches.

In our model, we advocate the minimization of *work* rather than the minimization of the ‘action’ to generate the equations of motion for a closed, conservative system. This is because we believe the term potential should technically only be used to represent the *potential to do work* rather than describing some mysterious, stored potential energy within the system of particles or objects. Thus, all interactions must minimize the change in kinetic energy of the objects involved. Most generally, and relativistically, this can be represented as

$$\delta \int (m(t) - m_0)\, dt = 0 \qquad (1)$$

or, with a little poetic license, we can write (where T is the kinetic energy)

$$\Sigma_i (\Delta T_i) = 0 \qquad (2)$$

Since all energy is fundamentally rotational kinetic energy, or can be represented as such, we can also write equation (1) as

$$\delta \int d\mathbf{L}/dt \bullet \omega dt = 0 \qquad (3)$$

Of course, it is not always necessary or desirable to eliminate the concept of potential, or potential energy, when solving physics problems. The important thing to remember is that potential energy is only a mathematical placeholder, shorthand, or ‘trick’, and should not be elevated to an ontological *ding an sich*.

With potential energy, the Crisis in Physics begins !

Introduction

Most people seem to agree there is a crisis in physics although they do not necessarily agree as to what the crisis is. Some believe the crisis is a lack of 'progress' since the invention of the standard model, and/or the incompatibility of the standard model with general relativity. Others believe that the standard model is a silly, incoherent, mystical mess. We happen to be of the latter camp. We cannot even say that it is bad metaphysics since there is no overarching, systematic structure or point of view, with the possible exception of the disaster that is gauge theory. Instead, we have a bunch of ad hoc assumptions all bunged together, including, but not limited to; collapsing wave functions, entanglement, nonlocality, gauge theory, the quantum mechanical vacuum, broken internal symmetries, disembodied fields, etc.

The one feature that the standard model and general relativity have in common is that they are both field theories. How fields became more than mathematical conveniences and were elevated to the status of entities permeating all space is both a mystery and a scandal! Originally, particles were the sources of fields, but now people seem to believe that particles are manifestations of the very fields themselves. Besides being almost impossible to conceive, these field theories have many 'infinities', some that seem inconvenient but ignorable, and some that are a cause for concern.

It is said that any new theory must be renormalizable. We believe that any theory that requires renormalization is no theory at all.

We believe the standard model and general relativity are incompatible because they are both way too complicated and defy common sense. They are more like self fulfilling prophecies than proper physics theories. There has been no progress in (theoretical) physics for the last hundred years.

There is no doubt that our still developing model contains some silly ideas that may not stand the test of time. However, it is not magical or mystical. It is logical, mechanical and physical, employing nothing more than the time honored notions of classical 'Newtonian' physics; mass, conservation of (relativistic) energy and momentum, and the concept of action and reaction.

Of course, the reader must judge for themselves! So, let us proceed !

We shall jump right into acceleration!

Acceleration

All acceleration involves the emission or absorption (or exchange) of photons. In our model, acceleration *is* the emission and absorption of photons. Hence, all acceleration results in a change of angular momentum of the accelerating particle in units of Planck's constant; h. Whether an electron is accelerating 'linearly' in an antenna or receiver, or making energy level transitions in an hydrogen atom, the acceleration is always discrete. All acceleration involves a change in angular momentum. *Acceleration is change in angular momentum.*

In our model, classical electromagnetic radiation, or the electromagnetic wave, is a result of the emission of many, many, (many!) discrete and individual photons from accelerated electrons. Each individual electron tries to resist acceleration (due to photon absorption!) by shedding energy and angular momentum. They'd rather remain "at rest."

The mass, or energy, of an electron can be completely described in terms of angular momentum as well, of course. We will explore this more later on.

In the classical picture of radiation, electrons that are oscillating at finite speeds emit waves, or cause wave disturbances, traveling at the speed of light. This discontinuity between the velocity of the electron and the resulting wave has not gone unnoticed by physicists. It is one of the many conundrums swept under the proverbial rug and saved for later resolution while people forge ahead unperturbed. Maybe string theory will solve this problem! We believe accelerated particles *literally shed* unwanted or excess energy and angular momentum in the form of the photon. These photons are not created from, nor do they pop out of, the vacuum. The photons are not stored or contained in the electron. Photons are unwanted or excess angular momentum ejected or released by accelerating particles.

When an electron makes an orbital or energy level transition, in an hydrogen atom for example, the electron emits one photon. The atom does not emit the photon, nor is the photon magically conjured up out of the vacuum. The electron sheds the photon as excess angular momentum. As our loyal reader knows, the energy and momentum of any and all particles is nothing more than the simple harmonic oscillation of the particle's intrinsic angular momentum or spin. Recently, "scientists have determined" that electron orbital transitions take a finite amount of time. This is because, the electron is literally accelerating, traversing space from one orbital to the next.

Spin is real. Photons are real. **Photons *are* spin.**

Photons are not created and destroyed. They are emitted and absorbed

Kinematics and dynamics

The Micheleson Morley experiment demonstrated the nonexistence of the ether. The ether was supposed to provide a medium in which light waves would oscillate and propagate. The ether was also supposed to provide a universal reference frame to essentially constrain the speed of light to a constant and well determined value. With this 'reference frame' abolished, the *hasty* conclusion was that all inertial observers, no matter their relative motion and velocities, would determine the same value for the speed of light. We say this conclusion is 'hasty' because it is too general and can never be empirically verified.

The implication is that two independent inertial observers would measure the same speed of a light ray or photon originating from a *separate and third* independent source! This, of course, is impossible. Measuring the speed of light is not like determining the relative speed of cars on the highway. One cannot measure the speed of light relative to local landmarks as it whizzes by. To measure the speed of light, the light source and the measuring apparatus must both be stationary and in the same inertial reference frame. Perhaps this is only a current and temporary restriction that may someday be overcome, but for now, it is true. Hence, all we can really assert is that all observers will measure the same value for the speed of light in their own inertial reference frames.

An observer in an inertial reference frame will measure an increase in the 'relativistic' mass of a particle or object if it is sufficiently accelerated. This determination is independent of any knowledge of, or reference to, any other existing reference frames, even though this phenomena came to light during the examination of the Lorentz transformations and the consequences of the postulates of special relativity. The acquisition of relativistic mass in an inertial reference frame can be derived without reference to any coordinate transformations as we attempted to prove in a previous paper.

The most egregious and popular misunderstanding of 'relativity' is that one can understand and determine the behavior of an accelerated object in an observer's inertial reference frame by transforming to the reference frame of the accelerated object and considering how it behaves in the frame where it is at rest. Of course, we are referring to the 'explanation' for the long lived nature of high energy cosmic ray muons in earth's upper atmosphere. We have dealt with this question (rather unsatisfactorily) in another paper and will not revisit it for now, at least.

The upshot is when you are doing physics and examining a particular phenomena, you must choose one particular reference frame and totally operate and calculate and speculate within it.

In our formulation of physics, *all* energy is kinetic energy. Potential energy is a convenient mathematical device, but potential energy should not be considered to be a real substance inherent in objects, systems, or fields. Fields cannot physically store energy or momentum or transmit said quantities. *Fields certainly cannot sustain stresses and strains.* Fields are not real. Particles are real.

In our model, particles have intrinsic mass and they actually spin. Photons *are* spin; or one unit of angular momentum. The projection of the photon's angular momentum, L = h, varies sinusoidally along the direction of travel with a frequency determined by its energy. More precisely, the energy of the photon is determined by the frequency of this oscillation! In the case of fermions (e.g., neutrinos and electrons), the angular momentum precesses about the direction of travel tracing out an 'isotropic cone'. This precession also results in a sinusoidal projection of the angular momentum along the direction of travel.

We have demonstrated (perhaps even correctly!) that the 'relativistic' formulation of the kinetic energy is $E = mv^2$. If we take a look at the familiar equation $E = mc^2$, for an electron, say, we can interpret this as the *kinetic energy of the electron.* In other words, the electron is *intrinsic* (relativistic) mass spinning at the speed of light. Without this interpretation, the formulation $E = mc^2$ has no real logical or physical significance.

The rest mass or the relativistic mass of the spinning electron is determined by the rate of oscillation of its angular momentum vector; a kind of spinning spin, if you will. This interpretation may strike the 'sophisticated' reader as naive. Compared to quantum field theory where electrons are barely things at all and don't spin (until necessary, of course!) and acquire mass through the Higgs mechanism, I'd say that it is a toss up at best as to which set of ideas are the more outlandish.

In our model mass is: inertial mass, gravitational mass, and *the only coupling charge.* The electric charge is a universal coupling constant. There is no bare electric charge, no charge screening, and the coupling constant, α, does not run. *The relativistic mass 'runs'.* For electromagnetic interactions we replace the electric charge, e, with

$$e \Rightarrow (e/m_0)m = (e/m_0)(m_0/\sqrt{1 - v^2/c^2}) = (e/\sqrt{1 - v^2/c^2}) \qquad (4)$$

If we insert this new definition of e into the Lorentz force, we see that Lorentz invariance arises in a completely 'natural' way.

$$\mathbf{F} = (e/m_0)m\,(\mathbf{E} + \mathbf{v}\times\mathbf{B}) \qquad (5)$$

Elementary particles

The photon is most likely the most elementary of the elementary particles, although it owes its existence to the acceleration of leptons! Regardless, let's begin our discussion by considering the electron.

In our mechanical model the electron may be considered a mathematical point particle, although we believe it to be extended in various ways. First, we take the electron to be a massive, spinning disk (rather than a blob or a sphere.) This spinning disk is just extended enough to allow the spinning mass to induce a classical electromagnetic field, or the electron's electric charge! We shall consider this idea of quantum mechanical electromagnetic induction in a later paper. The electron also has an operational, or effective, radius corresponding to its wavelength. Even at rest an electron has a wavelength, the Compton wavelength, since an electron is always spinning or 'precessing' about some preferred axis or direction. Our model of the electron spin and mode of propagation is depicted in Figure 1.

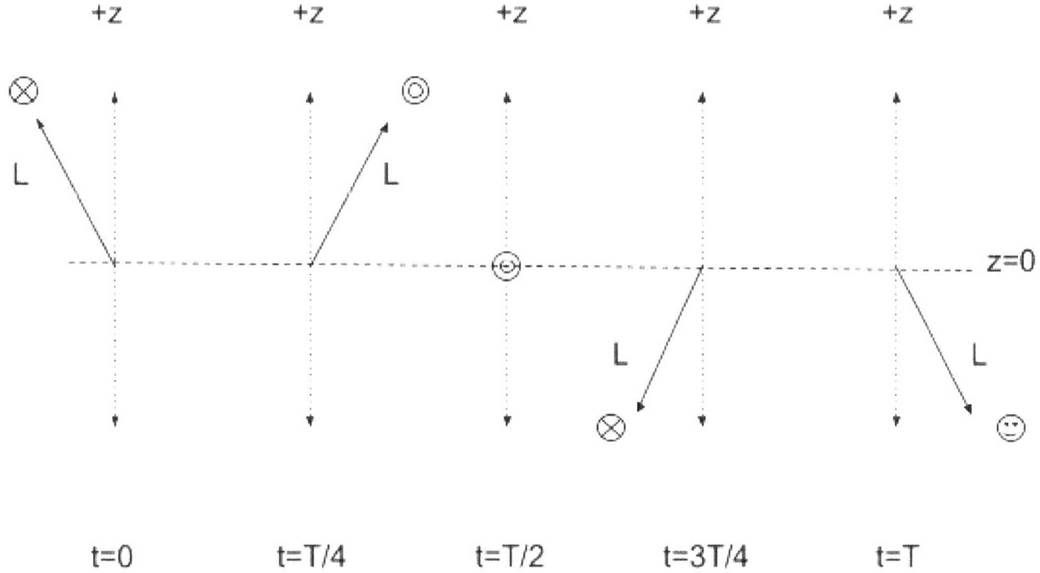

Figure 1: At rest with a lepton traveling in the z-direction. The spin angular momentum vector 'precesses' about the direction of motion, tracing out a closed, three dimensional figure eight. The x symbol represents motion into the page. The dot symbol represents motion out of the page. At time T/2, we see the angular momentum is *perpendicular* to the direction of travel. To represent an antilepton, simply swap the x symbols and the dot symbols.

As the electron propagates and precesses, the projection of the angular momentum along the direction of travel varies sinusoidally (although the polarization remains constant.) This can be inferred from Figure 1, and is illustrated schematically in Figure 2.

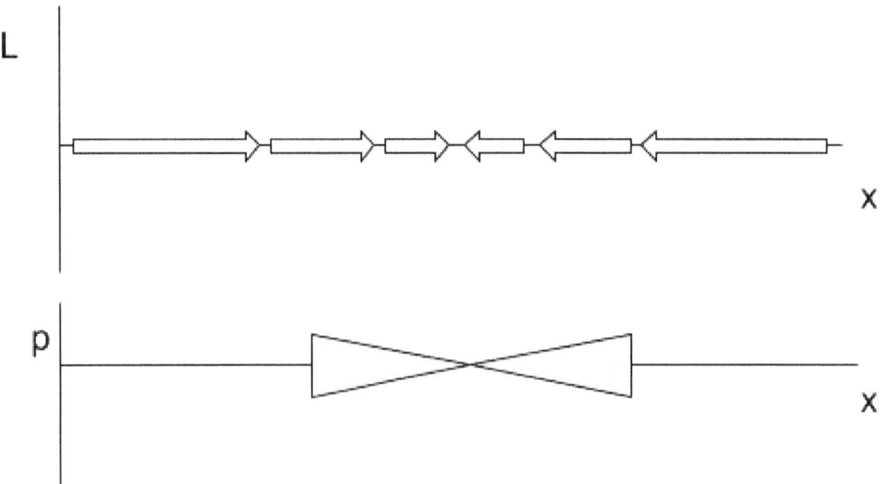

Figure 2: The top drawing depicts the sinusoidal projection of the angular momentum of an electron as it propagates. The bottom drawing depicts how the *linear momentum* 'available' for interaction with other particles, or an observer's measuring instrument!, varies with position (and time.)

The electron is only able to absorb or emit 'real' photons in proportion to the extent that the angular momentum is available for such interactions, or according to the percentage of the angular momentum projected along the axis of rotation. We remind the reader that for real photon absorption or emission, the angular momentum of the electron must change by +/- 1h, where h is Planck's constant or the angular momentum of the photon. When the angular momentum is *perpendicular to the direction of travel,* the electron is susceptible to and deals with 'virtual' photon interactions. Thus, for the canonical toy model of an electron in a box we can understand why the wave function, and the probability for measurement or detection, goes to zero at the 'nodes' as well as at the walls of the box.

From this simple toy model, we can understand the origin of the **Born rule**, or why the square of the wave function represents the probability for detecting or measuring a particle at a particular place and time. Figure 3 shows the probability distribution of the particle in a box for the energy state n = 2.

In our model, the wave function for a (free or bound) particle is *not* a plane wave of probability. The wave function represents the (not quite so) simple harmonic oscillation of the particle's angular momentum as it propagates in a well defined direction along a well defined path. A particle traveling along the 'x-axis' *cannot* be and is not simultaneously anywhere and/or everywhere in the x-y and y-z planes. A wave packet is not necessary to describe the behavior of a single particle but only the unavoidable spread in momentum and energy of an experimentally prepared beam or bunch of particles.

We do not dispute the mathematics of ordinary and relativistic quantum mechanics, only the crazy 'interpretation' of what it all means or signifies. Our **physical model** can explain the behavior of the ideal situation of a particle in a box, as well as the more realistic picture of particles bound by some finite potential well, and the phenomenon of quantum tunneling.

We *do* reject quantum field theory, although we retain the idea of the 'virtual' photon. The idea of the virtual photon is as metaphysical as our model gets, although the behavior of this virtual photon differs from that of quantum field theory in a number of ways. In our model, the virtual photon represents a constant and continuous connection between particles, involving the *constant* interchange of energy and momentum. No matter their separation, interacting particles *always* move 'in *tandem* and *simultaneously*'. Virtual photons are not emitted and absorbed. Virtual photos do not split into particle antiparticle pairs.

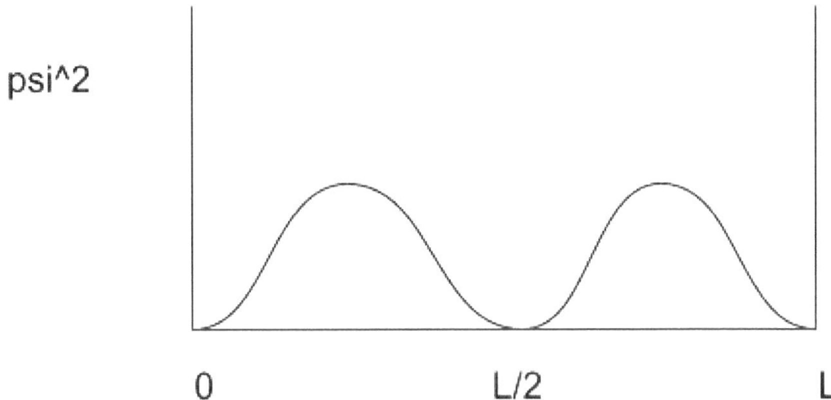

Figure 3: Probability distribution for a particle in a box; n = 2.

Now, just as an accelerated particle will shed energy and angular momentum in the form of the photon, sometimes high energy particles will shed angular momentum *and matter* in order to minimize the difference in the rest mass and kinetic energy of a particle or system of particles as described by equation (1). Enter the neutrino! We assume the neutrino to be the massive analogue of the photon. For example, a muon will shed excess mass energy and angular momentum in the form of two neutrinos resulting in an electron as illustrated in Figure 4. In this model, the muon really *is* a 'massive electron', and the difference between the energies of the three lepton families can be explained by an appeal to equation (1) as we have tried to demonstrate, in at least a hand waving fashion, in several previous papers. Similarly, the three lepton neutrinos differ *only* in their relative kinetic energies.

Clearly, our model of muon decay does not involve the emission, exchange, or decay of a *massive, charged, virtual gauge boson.* This model does not require any additional particles or fields nor the existence of the quantum vacuum.

Particles (leptons) *shed* unwanted, or excess, angular momentum in the form of the photon.

Muons and tau leptons *shed* excess matter and angular momentum in the form of the neutrino whenever the conditions (i.e., available 'phase space') allow.

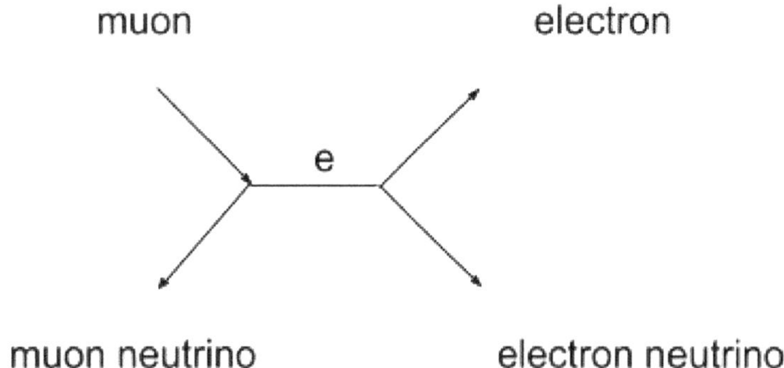

FIGURE 4: Muon decay. The 'propagator' is a 'virtual' electron.

During the formative years of modern physics, the newly discovered electron was determined to have mass, charge, *and* spin. It was also determined that such a tiny spinning mass and charge could not account for the observed angular momentum or magnetic moment of the electron. So it was decided that an electron must not actually spin even though it has a well defined 'classical' angular momentum and magnetic moment, and appear as these 'classical' quantities in our 'classical' detectors. Later, of course, the electron would lose its intrinsic mass to the Higgs mechanism. The electric charge of the electron was deemed to be A Consequence Of Electromagnetic Gauge Invariance.

In our conception of physics, angular momentum is the fundamental unit or building block of Nature, Reality, The World -- let's say. All Physical Objects.

Rather than deriving the concept of angular momentum from spinning mass, we have derived the concepts of mass and spin from angular momentum! We argue that *all* particle properties: mass, spin, angular momentum, linear momentum, and *even electric charge* are manifestations of this fundamental angular momentum.

The simplest physical object in the universe is the photon. It is one unit of angular momentum. In our model the 'virtual' photon is also the mediator of the gravitational and electromagnetic forces. Using these two notions, hindsight, and a bit of teleology perhaps, we shall try to construct the whole world from scratch.

We have discussed the fundamental nature of photons and the leptons and their mode of propagation earlier in this paper and *ad nauseum*, elsewhere. The interesting thing to note for the photon is the energy of the photon is the rate of change of the angular momentum vector along the 'z axis' or direction of propagation

$$L = h \; ; \qquad \langle L_z \rangle = h \; ; \qquad E = dL_z/dt = h^{bar} \upsilon \qquad (6)$$

$$L_z(t) = h \exp(-i\, 2\pi\, \upsilon t) \quad ; \quad E(t) = -i\, dL_z(t)/dt \quad ; \quad z(t) = ct \qquad (7)$$

Equation (7) represents the 'reverse pilot wave' representation of the photon.

The photon **is** this angular momentum vector (a 2-D point particle?) tumbling or spinning or spiraling in three dimensional space.

The photon has energy and mass, but is not considered **matter.**

Our definition of matter will be that $|L| > |L_z|$, and that this matter can emit and absorb our photons. We know that this matter, the lepton, has $L_z = h^{bar}/2$ (the "spin") and $L=\sqrt{3}/2\ h^{bar}$. The simplest lepton is the 'electron' neutrino. We now believe the neutrino to be (rest) massless. This essentially means there is no lower bound on the frequency at which the angular momentum vector of the neutrino can precess about the direction of travel. This implies there is no upper bound either, so our neutrino could be quite massive, if necessary.

In our model, the three spin components of the lepton, $s_x = s_y = s_z = h^{bar}/2$, are *linearly independent*; the only constraint being that they yield $L=\sqrt{3}/2\ h^{bar}$.

Our definition of the electron will be a lepton which is *constrained* to have a *lower bound* on its frequency of oscillation. This lower bound then determines the rest mass of the electron; $m_e = (h^{bar}\upsilon_e) / c^2$, and also, essentially, the *nature of electric charge*.

What fixes this lower bound is currently a mystery. Perhaps we must take it as a given as we do the value of Planck's constant or the speed of light.

How this angular momentum vector, precessing at a constant rate, generates the particle mass; *and* the magnetic moment; *and* the electric charge is all explained by our bootstrapping model called Quantum Mechanical Electromagnetic Induction. For now all we would like to say about this mechanism is that the electric charge is associated with *'ponderable' mass.* The neutrino would then be considered 'imponderable' mass.

Let's recap the situation so far. We started with the simplest vector particle, the photon, as a known or given; with angular momentum, $L = h$, and velocity, $v = c$. We then constructed the simplest *'spinor' particle* that could emit and absorb this photon; the lepton (i.e. the neutrino and the electron, etc.) We have already discussed the spinor nature of the lepton here and in greater detail elsewhere. The important point to note is the energy of the lepton is $h^{bar}\upsilon$!

The energy of the photon is $dL_z/dt = h^{bar}\upsilon$. The energy of the lepton is $dL_z/dt = h^{bar}\upsilon$. How can this be if $L_z = h^{bar}/2$ for the lepton!? This is because the angular momentum vector of a *spinor* precesses twice around the axis of travel per every one period of oscillation.

The energy of an electron is a combination of rest mass energy and kinetic energy, T

$$E = mc^2 = (m_0 + m_T)c^2 \qquad (8)$$

$$E = h^{bar}\upsilon = h^{bar}(\upsilon_0 + \upsilon_T) = dL_0/dt + dL_T/dt \qquad (9)$$

Gravity

Our idea of gravity, 'quantum', quantum mechanical, or otherwise, is not that it is similar to the Coulomb potential (which was originally inspired by gravity, of course), but that it is essentially identical with the electromagnetic force, except for the 'coupling constants'.

We can construct a Lorentz force for a particle in a gravitational field

$$\mathbf{F_g} = m_1 (\mathbf{E_g} + \mathbf{v}x\mathbf{B_g}) \qquad (10)$$

where $E_g = m_2G/R^2$, and $B_g = (G/c^2)(\mathbf{p_2xR})/R^3$. The cross product of equation (10), $\mathbf{p_1x(p_2xR)}$, 'mixes' the space and momentum components of the two massive objects. In our model, this relative angular momentum of the one particle to the other contributes to the total mass difference between the two particles, and hence to the overall gravitational charge.
For circular orbits, where $m_1 = m_2$, $p_1 = p_2 = p$, the cross product term becomes $2p/R^2$.

The 'universal' gravitational force for the interaction between two bodies, normalized by the total energy of the system, includes this new cross product factor and can be written

$$\mathbf{F}/E_{TOT} = K^*(c/R)^2\mu - K^*(\mu v^2/R^2) - K^*(c^2 l^2/\mu R^3) \qquad (11)$$
$$K = G/c^2 \qquad (12)$$

where E_{TOT} is the sum of the relativistic masses of the two particles (a misnomer!), μ is the reduced mass, and v is the (relativistic) velocity of the reduced mass. All masses are, or at least can be, relativistic. In our model, relativistic mass is a 'classical' quantity. We have constructed this new universal gravitational relationship by incorporating our new coriolis force and 'relativistic mass' into the conventional two body gravitational force equation, but requiring this pesky new normalization. Someone should be able to turn this into a tractable and conventional Lagrangin or potential energy equation, and maybe even solve it!

Probably not me.

In our model, the gravitational interaction is mediated by the virtual photon. Each virtual photon, connecting each individual particle pair, is concerned solely with the relative angular momentum and the relative rotational kinetic energy of the two particles.

The hydrogen atom

Our thoughts on the hydrogen atom are similar. The cross product, or coriolis term, should be included in the Hamiltonian or Lagrangian of the system along with the *relativistic* masses of the particles (at least formally!) from the get go, rather than being put in later as corrections by hand. The velocity of the electron should be taken seriously and considered to be real, or true, or actual, or however you may like to express it. All higher order energy corrections would then stem from a series expansion of the relativistic mass(es) in terms of v/c.

The hydrogen atom is obviously *not* an electron in a static Coulomb potential.

We know that the fine structure constant can be defined or construed to be the velocity of the electron of the hydrogen atom in the lowest orbital state relative to the speed of light

$$\alpha = v_1/c \qquad (13)$$

which means we can write the relativistic mass of this electron as

$$m_1 = m_0 /(1 - \alpha^2)^{1/2} \qquad (14)$$

Probably no great revelations here, but my tiny mind is exploding.

Interaction

It seems in many ways the more natural variables for discussing elementary particles are the wave numbers (k,ω) rather than the momentum and energy, (p,E); particularly as we are concerned with *casting all of physics in terms of angular momentum conservation.* We have suggested elsewhere that the favored canonical variables should not be (x,x^{dot}), but rather (θ,θ^{dot}) ! Then we could consider the wave numbers (k,ω) as representing the angular variations $(d\theta/ dx , d\theta/ dt)$ of the oscillating angular momentum vector, L, along the direction of travel, during the time of travel or time of interaction.

In a very simple, <u>very</u> hand wavy, example of the elastic scattering of two particles, we would then have three wave functions. One wave function for each of the two 'real' particles, and one for the virtual photon;

$$\exp(-i(k_1r_1-\omega_1t)) \ ; \ \exp(-i(k_2r_2-\omega_2t)) \ ; \ \exp(-i(\Delta k(r_1-r_2) -\Delta\omega t))$$

Now, if we know $\mathbf{k}_1$, $\mathbf{k}_2$, and $\Delta\mathbf{k}$ at time, $t=t_0$, and the coupling constant (e.g., G), then we essentially know the energy and momentum of the two particles and also the quantity $1/R^2$ at time $t=t_0$. Next, we naively suggest that any interaction will minimize the change in frequency of the virtual photon, $\Delta\omega$, for some given force or potential or set of constraints. So easy!

All interaction occurs over some period of time between two or more particles in some region of space. We stress that physics should be concerned solely with these interactions rather than instantaneous 'events'. Uncorrelated 'events' considered in different reference frames opens up a whole can of worms concerning simultaneity, and whatnot. As we know, in cases of obvious cause and effect, or interaction, there is no question of which event precedes which, in any reference frame.

All interaction deemed 'action at a distance' involves the constant exchange of energy and momentum via a single virtual photon coupling any two particles. For planets, we can consider them to be point particles tethered (to the sun) by a single virtual photon after all the appropriate integrations. These virtual interactions, or interactions at a distance, *cannot change the overall angular momentum of any system of particles* even when the 'vector cross product' is involved in the equations of motion. We believe this conservation of angular momentum is often neglected or overlooked when considering problems with fixed mass or fixed charge centers or situations involving fixed, external fields. Take as an example; the electromagnetic induction of current in a wire coil by moving a permanent magnet through it. What we really have are many tiny electrical currents in the magnet tethered by virtual photons to electrons in the wire coil. Relative motion between the two then induces current in the wire in the 'usual way'.

There can only be a change in angular momentum, when there is *real photon emission or absorption,* and these photons must 'leave or enter' the closed system.

Ultimately, photons couple to the angular momentum, or mass, of a particle. A particle absorbing a real photon will experience an increase in angular momentum, either in its orbit, or, as an increase in the rate of precession of the angular momentum vector of the particle along the direction of propagation.

The *electric charge* represents a fixed lower bound on the rate and *direction* of the precession of angular momentum. (Electrons spin to the left, positrons (and protons!) spin to the right.) The *strength* of the electric charge, or electromagnetic force, is due to this stubborn excess spin or angular momentum. Charge in aggregate (and seen from afar) produces neutral mass. ***So***, the electric coupling strength represents, in large, an interaction between the rotational angular momentum of any two particles, and the gravitational coupling strength represents the interaction between their rotational kinetic energies.

Whether 'solving equations', determining the coupling strengths of the "four" forces, or determining the magnetic moments of the leptons, all of our series expansions are in terms of v/c. *The electric charge*, e, *is strictly constant*. Even so, I would wager that the values in our expansions, if done, would match those of QED, term by term, for several quantities.

In our model particles are not created or destroyed, rather, particles interact, combine, and decay by exchanging, emitting, and absorbing mass (the photon) or matter (the neutrino). Rather than creation and annihilation operators, we need something akin to spin step up and step down functions.

There is still much work to be done on this awesome model, people.

General Relativity represents physics by celebrity. The Copenhagen Interpretation of quantum mechanics represents physics by bullying, and the Standard Model physics by committee. It is said that no one person can resolve the discrepancies between, or unite, these 'great pillars' of physics because the problem is too big and too complicated.

I agree. However, one person may be able to point out that things have gotten way too crazy. It is time for a complete reevaluation of modern physics, particularly from an ontological and epistemological perspective.

Conclusion

Lest this book be no more than a selective rehash of what we have said or tried to show before in previous books, we beg the reader's indulgence and claim we are building our model logically and synthetically from the ground up, again. In the process we have refined or corrected several previous ideas and discovered several new ideas we must now insist upon as our physical model of particles and particle interaction continues to evolve.

The first correction to our model is the neutrino now has no rest mass. The neutrino is considered to be a matter particle with no lower bound on the frequency of oscillation, or rate of precession, of the angular momentum $L=\sqrt{3}/2\ h^{bar}$ along the direction of travel. We decided to call this imponderable mass although the relativistic mass of the neutrino may be quite large. The electron is the first instance of ponderable mass. The electron has a rest mass, or *rest frequency*, which is fixed. Furthermore, this fixity is represented or demarcated somehow by or as the electric charge through the wonderful and still currently mysterious mechanism of quantum electromotive force which we have decided not to pursue further quite yet. If the muon did not have the same electric charge but a different rest frequency from the electron, we could just say that's how things are and call it a day! It seems that the 'mystery' of Quantum Mechanical Electromagnetic Induction and the order of the Three Particle Families are intimately tied together and we expect to investigate this relationship soon in a later book. For now, all we'd like to say is the rest mass of the electron designates energy that cannot be mechanically dissipated away.

The first new rule that we uncovered, or now postulate, is that the energy of *any* particle, be it a photon, a neutrino, or an electron, is equal to the rate of change of the component of the 'internal' angular momentum of the particle along the direction of travel; $E = dL_x/dt = h^{bar}\upsilon$. Any additional angular momentum a particle has relative to a second particle will contribute to the overall relativistic mass. Up to now, we have been using the z-axis as the canonical direction of travel for (free) particles, as we often consider them 'at rest', but now realize the x-axis is the better choice in most circumstances as we shall see in just a moment.

Our second new rule, or hypothesis, is that the three components of the internal spin of the electron (or any lepton), s_x, s_y, s_z, are *linearly independent*. Perhaps this is nothing new. In our previous thinking, we had only one 'free' component of angular momentum, $h^{bar}/2$, and this was projected sinusoidally along the direction of travel. *This propagation mechanism determines the projection, or direction, of one component of the particle spin*, before the introduction of any external or internal magnetic fields. This model does not leave any extra free spin to align with a magnetic, $\mathbf{B_z}$, for an electron travelling in the x-y plane, for example. So we must assume *three independent spin components*; two of which are constrained and can be determined by us in the example cited just above! The third component of spin is also then constrained, but unknowable by us until measured.

Books by Greg Feild

1. “A quantum mechanical theory of gravitational interactions”
   CreateSpace Independent Publishing, 8/29/2016

2. “Observations on the quantum mechanical nature of gravity”
   CreateSpace Independent Publishing, 10/8/2016

3. “On gravitation and electric charge”
   CreateSpace Independent Publishing, 10/29/2016

4. “On spin, mass, and charge”
   CreateSpace Independent Publishing, 11/29/2016

5. “On angular momentum, acceleration, and absolute motion”
   CreateSpace Independent Publishing, 1/1/2017

6. ”The Sinister Universe”
   CreateSpace Independent Publishing, 3/1/2017

7. ”On Parity and Isospin”
   CreateSpace Independent Publishing, 4/11/2017

8. “Reflections on the Sinister Universe”
   CreateSpace Independent Publishing, 5/12/2017

9. “On Current Physics”
   CreateSpace Independent Publishing, 6/11/2017

10. “A Critical Examination of Classical and Quantum Mechanical Waves”
    CreateSpace Independent Publishing, 6/18/2017

11. “On wave particle duality and the quantum of action”
    CreateSpace Independent Publishing, 7/6/2017

12. “On matter, mass, and motion”
    CreateSpace Independent Publishing, 9/14/2017

13. “On action and reaction”
    CreateSpace Independent Publishing, 9/24/2017

14. “A quantum mechanical theory of everything”
    CreateSpace Independent Publishing, 11/5/2017

15. “On Interaction”
CreateSpace Independent Publishing, 4/21/2018

16. “On Rotation”
CreateSpace Independent Publishing 8/19/2018

17. “Revenge of the Sinister Universe: The Reality of Everything’
CreateSpace Independent Publishing, 9/4/2018

18. “On Math, Physics, and Metaphysics”
CreateSpace Independent Publishing, 10/1/2018

19. “On Quantum Mechanics”
CreateSpace Independent Publishing, 10/15/2018

20. “On Epistemology and Ontology”
CreateSpace Independent Publishing, 10/21/2018

21. ”Toward a Metaphysics of Mass and Motion”
CreateSpace Independent Publishing, 10/29/2018

Compilations:

A. “The Universal Model of Our Sinister Universe: The First Ten Books”
CreateSpace Independent Publishing, 7/2/2017

B. “The Canons of the Sinister Universe:
The Last Four Books on the Universal Model of Our World”
CreateSpace Independent Publishing, 11/5/2017

C. “The Return of the Sinister Universe: The Immaculate Collection”
CreateSpace Independent Publishing, 9/4/2018

D. “The Battle for the Sinister Universe: The Heuristics”
CreateSpace Independent Publishing, 10/29/2018

22. “Postmodern Physics”
Kindle Direct Publishing, 10/28/2020

Resources

Quantum Field Theory
Claude Itzykson, Jean-Bernard Zuber

Atomic and Quantum Physics
H. Haken, H.C. Wolf

Modern Elementary Particle Physics
Gordon Kane

Classical Dynamics of Particles and Systems
Jerry B. Marion

Foundations of Electromagnetic Theory
John R. Reitz, Frederick J. Milford, Robert W. Christy

Quantum Physics
Rolf G. Winter

Gauge Theories in Particle Physics
I. J. R. Aitchison and A. J. G. Hey

Quarks and Leptons: An Introductory Course in Modern Particle Physics
Francis Halzen, Alan D. Martin

Quantum Field Theory
F. Mandl, G. Shaw

Theoretical Mechanics of Particles and Continua
Alexander L. Fetter, John Dirk Walecka

The Theory of Spinors
Elie Cartan

Elementary Modern Physics
Richard T. Weidner, Robert L. Sells

Quantum Mechanics
Claude Cohen-Tannoudji, Bernard Diu, Franck Laloe

Deconstructing Modern Physics

Greg Feild

January 28, 2022

About the author

I earned a Ph.D in experimental high energy physics from the Pennsylvania State University working on HERA at DESY in Hamburg, Germany studying photoproduction and deep inelastic scattering in electron-proton collisions.

I did my postdoctoral studies with Yale University working at Fermilab on the CDF experiment at the Tevatron. My primary research interest was particle hadronization in quarkonium production in proton-antiproton collisions.

I have always found the standard model unsatisfactory, even when I had less of an understanding of it than I suppose myself now to have! These misgivings about the standard model are quite distinct and separate from my disdain for the Copenhagen interpretation of quantum mechanics. Putting all mathematical complexity aside, these theories are conceptually incoherent and rationally inexplicable. To claim this is just how the world works is merely begging the question and no answer to skeptical inquiries.

Abstract

The field of modern theoretical physics is in need of a major course correction. In this book, we argue that current physical models and physics theories have way too many unobservables, way too many forces, way too many fields, and just way too many variables to function as a proper theory at all.

Forget about revolutions and paradigm shifts. Revolutions are never really more than surface ripples. To save physics, we need to keep shoring up the original foundation while removing the hideous, gilded, baroque facade it has acquired over the last one hundred years. Thus, we offer a new metaphor and a new mechanically motivated model for a new millennium.

an ahistorical approach

Introduction

Should we even speak of modern physics anymore? Most of the maladies of modern physics have been inherited from classical theory, where they were similarly problematic. I am thinking in particular about field theory, of course, and the cascading and ever more far-fetched and abstruse abstractions it has spawned.

Do we really share the astonishment of the physicists of the turn of the last century at the mysterious aspects of the behavior of elementary particles, or is such an attitude merely (sincerely) feigned? Our astonishment should be that such talk still abounds. Is such an attitude still constructive for doing physics?

Can we reign in current theories and speculators? Can we still accept explanations that are really no explanations at all? Can we not with hindsight, and foresight, begin again and build a better physical and metaphysical model of particles and their interactions? I'd like to think that we can and that we are well underway.

My main gripe with *all* of physics (classical and quantum mechanical; general relativity, solid state physics, etc.) is the *reification of fields*. People believe, or must operate as if they believe, that fields are real actual things (items, objects, waves?) permeating all time and space and responsible for the behavior of the microcosm as well as the macrocosm. Some people are quite serious in this belief.

As everyone knows, fields are a convenient calculational and experimental device, derived from the original laws governing the forces between two massive or charged bodies. Once these fields became more than just mathematical placeholders for the force that a test charge would feel at a particular point in space and time, it became requisite that these fields store energy, suffer stress and strain, etc., and worst of all, particles began interacting with themselves!

Particles do not interact with themselves. Fields were not meant for such close up work.

At the next level of abstraction from the (vector) field we have the scalar potential field; arguably much more useful for many physical problems and perhaps more beautiful, but more dangerous as well. Are we not already on the slippery gradient? From the scalar and vector potentials, we then can construct the four-potential; a concept we quite like, actually, and which we have reappropriated for the foundation of our revised theory. For us it is an inflexion point between the two extremes of fields and gauges. Gauge transformations, like fields, are mathematical devices designed to make certain physical calculations or mathematical representations easier or more compact, etc. From these humble beginnings they have become generators of all sorts of charges and forces. People like to say Nature "knows" about gauge invariance !

Nature does not know about gauge invariance. Nature doesn't know about groups.

Nature does not know about symmetry; hidden, broken, or otherwise. The only symmetries in our model are due to translational and rotational invariances. And parity reflection, probably.

I believe all symmetries uncovered by the standard model are mere inventions; mistakes due to invalid fundamental assumptions, theories constructed to keep these assumptions alive. Nature *is* symmetric in our theory. How could it be otherwise? For every action (appropriately defined) there is an equal and opposite reaction. If there is a spin to the left, there is a spin to the right, etc. Current notions about parity violation stem from misunderstandings about the nature of particle spin, helicity, and how to *assign a parity* to a particular particle. I believe we have shown this sufficiently in previous papers, but shall discuss some of these issues more in this book.

The nature of spin is the keystone of our new model. In our interpretation of quantum mechanics, particles don't spin, or "have" spin; particles *are* spin. We have proposed a mechanical model of spin, that is not quite 'classical' (we're dealing with spinors, after all), but one which appeals to both the human imagination and intuition, and is useful for explaining physical phenomena both qualitatively, and quantitatively as well. From the spin we can derive the particle mass, energy, angular momentum, and most likely the charge.

In our model, an electron, for example, has three linearly independent spin components; $s_x = s_y = s_z = h^{bar}/2$. The direction and orientation of these three spin components are fixed and constant and continuous and are maintained during a particle's motion discounting interactions with other particles, of course. In principle (and perhaps only ideally), we can determine two of these three components of a particle's spin; one from the particle helicity (the projection of the spin s_x along the direction of travel), and one from the particle's behavior in a magnetic field (the projection of the spin s_z 'up or down'). This then leaves the spin component s_y free to point 'up or down' according to some *initial conditions.* This spin orientation is unknown only to us, hence the appearance of complex numbers in the spin matrix σ_y. The spin system of equations is mathematically *underdetermined* but *the physical system is completely determined* (and known 'to itself' !).

Electrons are 'matter' and spin to the left. Due to the spinor nature of the electron it *always spins to the left,* no matter whether it is pointing up or down. Its helicity is always left-handed. Positrons spin to the right. When an electron and a positron meet or collide, they combine to form a virtual photon in a completely 'natural and physical way'. There is no magical annihilation or burst of energy, no vacuum, no conjuring of photons. Only particle combinations and decays.

In our model, *this spinning to the left is negative charge*. As an electron is accelerated, its angular momentum or rate of spin to the left increases, and technically one could say that its electric charge increases, as is well known. The reason why the electron has a fixed 'rest charge' is that there is a fundamental lower limit on the frequency of the electron 'spin'.

The origin and mechanical nature of the lower bound on this rate of spin is one of the mysteries of our model. The next mystery, of course, is what fixes the lower frequencies of the muon and tau in such a manner that their rest charges are the same as that of the electron.

The neutrino differs from the electron only in that there is no lower bound on the frequency of its rate of spin. Hence, the neutrino is a rest-massless matter particle. The photon is a rest-massless mass particle. The difference between mass and 'matter' is the difference in the spin projections and modes of propagation. We have discussed this previously and shall not pursue it further here.

Now is as good a time as any to introduce the Metaphysic we have developed for our model over the last several books. Each particle, or system of particles, seeks to be in their ground state, or lowest possible energy state given the circumstances. (One might say all particles want to be 'at rest'.) This is essentially the same idea underlying current classical and modern physics with the minimization of the action given an appropriate Langrangian. In our new model, we endorse a similar approach, but with the minimization of work. We try to avoid the concept of potential energy (and forces, if possible), focusing instead on the kinetic energy of the particle or system of particles under consideration.

But, back to our Metaphysic. Each particle is connected to every other particle via a virtual photon, hence our retention of the four-potential from the standard model. However, this four-vector represents the energy and momentum, or 'wave function', of the virtual photon connecting the two particles, shuttling energy, momentum, and angular momentum back and forth. Each particle considers every other particle as a possible sink for its excess energy, and vice versa, of course! Two electrons spinning (by definition) in the same direction will each try to dump their excess angular momentum or rotational kinetic energy to the other. This constant contest or perpetual exchange will accelerate or drive the electrons apart due to conservation of energy and angular momentum and the explicit nature of our new model of electron propagation. In the case of an electron and a positron, the mutual exchange of opposite angular momenta will cause the two particles to move toward one another. (We shall discuss gravity in a moment.)

Consider a two particle universe consisting of an electron and a positron for purposes of conservation! The two particles interact via one virtual photon. There is no need to fill the universe with fields or infinite amounts of energy.

The upshot of this discussion is that *angular momentum* or *mass* is the fundamental *universal charge*. Electric charge makes its appearance as a fundamental, irreducible, constant spinning to the left; the rate dictated by the rest mass of the electron.

Clumps of neutral matter, separated at a distance, no longer transfer any net angular momentum via the two 'transverse' channels of the virtual photon leaving only the spacelike or longitudinal channels for net momentum transfer, and these two channels are inherently attractive. This is our theory of gravity.

It must be remembered that mass is derived from and/or is reducible to angular momentum. This includes intrinsic rotational angular momentum as well as relative angular momentum. So the gravitational force between two massive bodies is dictated by the total mass-energy difference, or relative angular momentum between the two bodies, and this includes the rotation of each body and the *relative motion* between the two bodies. Hence our expression of the gravitational force is a direct analog of the equation for the electromagnetic Lorentz force. In fact, in our model, the gravitational and electromagnetic forces are completely equivalent in their behavior and mathematical expression; only the strengths of the coupling differ.

So, rather than trying to unite quantum mechanics and general relativity, we decided general relativity had to go. In our model all gravitational interaction is mediated by virtual photons. *All* interaction is mediated by virtual photons. There are no massive gauge bosons. There are no leptonic propagators; hence in our model the photon must be a self-coupling gauge boson. This is necessary to account for the Compton effect as well as the bending of starlight by the sun!

In addition, there are only two forces; gravity and electromagnetism. The weak and strong forces represent the manifestation of the two extremes of *relativistic gravity* at the subatomic level. Call it Quantum Gravity if you must! There are no quarks or gluons, no gravitons, no Higgs boson. There are only leptons and photons.

This is our new model in a nutshell. All of these ideas have been presented to varying degrees in previous publications. It would be 'impossible' and redundant to reproduce all the arguments here. However, we would like to look at the fundamental nature of elementary particles and their modes of propagation one last time since this is the bedrock of our theory.

After a brief overview of the nature of elementary particles and how they interact (in our model), we will give a report on the status of two important aspects of our model that need more development, namely; the phenomenon of quantum mechanical electromotive force and the origin and nature of the particle families. Finally, we shall make some cautious statements about time and space.

Elementary particles

As stated in the introduction, in our model an electron (or any lepton) has three linearly independent spin components; $s_x = s_y = s_z = h^{bar}/2$. An electron *is* the vector $L = \sqrt{3}/2\ h^{bar}$. This vector 'precesses' (in a special spinor way) at a minimum rate determined by the electron rest mass. As the electron propagates and precesses, the projection of the angular momentum along the direction of travel varies sinusoidally (although the polarization remains constant.)

Our model of electron spin and mode of propagation is depicted in Figure 1.

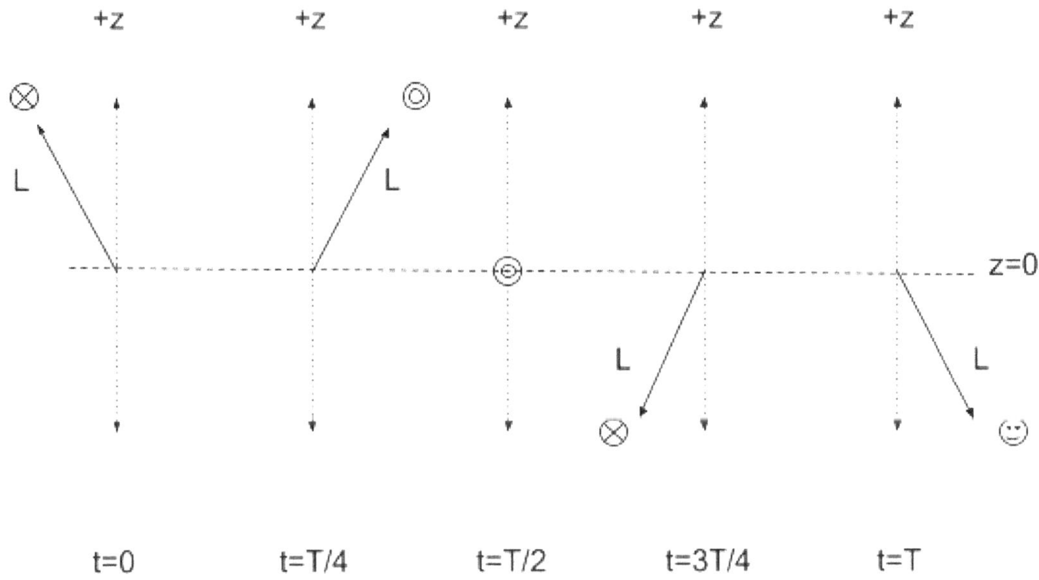

Figure 1: At rest with a lepton traveling in the z-direction. The spin angular momentum vector 'precesses' about the direction of motion, tracing out a closed, three dimensional figure eight. The x symbol represents motion into the page. The dot symbol represents motion out of the page. At time T/2, we see the angular momentum is *perpendicular* to the direction of travel. To represent an antilepton, simply swap the x symbols and the dot symbols.

Referring to Figure 1, we see that at time T/2 there is no projection of the angular momentum along the direction of travel, hence an electron is unable to absorb or *interact with a real photon at these sorts of nodes*. These nodes are where the particle catches up with its virtual interactions and/or bounces off a container wall!

We see from Figure 1 that the angular momentum vector 'circles' the direction of propagation twice before returning to its original position. Thus, rather than a point particle, the electron is a special simple harmonic oscillator, or a spinor.

The electron has an operational, or effective, radius corresponding to its wavelength. Even at rest an electron has a wavelength, the Compton wavelength, since an electron is always spinning or 'precessing' about some preferred axis or direction. In our model we want to express everything in terms of rotation, so we have recast, derived, or reimagined the mass and energy of the electron as or in terms of moments of inertia and angular frequencies. A summary is presented below.

In our model the angular momentum and the energy of an electron can be written

$$L = \sqrt{3}h^{bar}/2 + I\omega \quad (1)$$

$$E = h^{bar}\omega_0 + I\omega^2 \quad (2)$$

$$\omega_0 = 2\pi c/\lambda_0 \quad (3)$$

where λ_0 is the Compton wavelength of the electron, which we also take to be the 'rest radius'.

We define the velocity dependent moment of inertia $I(\lambda)$ to be

$$I = m\lambda^2/(2\pi)^2 \quad (4)$$

From equations (1) and (2) one can see that the 'rest angular momentum' and 'rest rotational kinetic energy' are given by

$$I_0\omega_0 = \sqrt{3}/2\ h^{bar} \quad (5)$$

$$h^{bar}\omega_0 = m_0c^2 = I_0\omega_0^2 \quad (6)$$

In our new model, the magnetic moment of the electron is thus velocity dependent (6,13,14,18);

$$\mu = (e/m_e)(h^{bar}/2c)(1 + \tfrac{1}{2}v^2/c^2 + \tfrac{3}{8}v^4/c^4 + \dots) \quad (7)$$

And we suggest the gravitational magnetic moment of the neutrino is

$$\mu = (h^{bar}/2c)(1 + \tfrac{1}{2}v^2/c^2 + \tfrac{3}{8}v^4/c^4 + \dots) \quad (8)$$

In our model the proton consists of two positrons and one electron bound by an interplay of the electromagnetic and relativistic gravitational forces. One can easily imagine the two positrons (one spin up and one spin down) forming a closed orbital shell around the electron nucleus.

A hydrogen atom would then contain *two electrons and two positrons*. There is no missing antimatter and no matter-antimatter imbalance in our model.

Here is a summary of the coupling strengths for the 'four forces' in our model.

The electromagnetic coupling 'constant', alpha, is (14);

$$\alpha = \alpha_0(1 + (v/c)^2 + (v/c)^4 + \dots) \qquad (9)$$

$$\alpha_0 = e^2/4\pi\varepsilon h^{bar}c \qquad (10)$$

The gravitational coupling constant is

$$\alpha_G = (m_e{}^2G)/(h^{bar}c)\ (1 + (v/c)^2 + (v/c)^4 + \dots) \qquad (11)$$

The weak coupling constant is

$$\alpha_W = (m_\nu{}^2G)/(h^{bar}c)\ (1 + (v/c)^2 + (v/c)^4 + \dots) \qquad (12)$$

And finally, the strong coupling constant (14) is

$$\alpha_S = (G/4\pi\varepsilon)^{1/2}\ (2m_e e/h^{bar}c)\ (1 + \tfrac{1}{2} v^2/c^2 + \tfrac{3}{8} v^4/c^4 + \dots) \qquad (13)$$

In the universal model, all the 'constants' run, because *the fundamental coupling charge* of a particle *is the relativistic mass-energy* of the particle.

The choice of the rest mass particles appearing in the coupling constants is conventional. In our model the neutrino no longer even has a rest mass. Perhaps some practical lower limit could be imposed.

In our model for two body interactions, only the 'lowest order diagrams', and only those involving 'one photon exchange' would contribute to the matrix elements. All higher order corrections would be due to the expansions of the particle masses in terms of v/c. Otherwise, the calculation of scattering cross sections, etc., would proceed much as they do now.

Will this approach yield sensible results agreeing with experiments? I don't know !

In our model the formulation of the Dirac equation is

$$H \psi = c\alpha \bullet p \psi \tag{14}$$

$$i\hbar \partial\psi/\partial t = -ic\hbar\alpha \bullet \nabla\psi \tag{15}$$

and is satisfied by, or solved with, the universal wave function

$$\psi = \exp(i(p\bullet x - (m-m_0)c^2t)/\hbar) \tag{16}$$

We can see this wave function contains the rest mass (or the global charge) and the relativistic mass (or the local charge). This is what one might call 'gauge invariance'. Certainly under Lorentz transformations the rest mass is constant and the relativistic mass is, well, relative.

This universal wave function describes the behavior of a spinor as depicted in Figure 1. This spinor satisfies the Dirac equation with two *positive energy solutions spinning to the left* and two *positive energy solutions spinning to the right*. We have real particles doing real things that we can at least conceptually picture and understand using mechanical models based on experience.

There are no particles traveling backwards in time, no 'holes' in the electron sea, etc. We have to pause and wonder at the origin of such absurd beliefs, why they were so readily adopted, and why people still fundamentally believe the same thing today, although in a more supposedly "sophisticated" way, of course. There seems to be no foundation for them, no ontological or epistemological basis or justification for making such explicit declarations. One cannot appeal to Pragmatism to support such grand Metaphysical claims.

Another feature of our formulation of the wave function is that it removes the explicit rest mass term from the Dirac equation; a term which has caused *all sorts of problems* for the standard model as far as I can tell. The ultimate nonsensical, paradoxical result is that any 'complete' theory should somehow account for the 'origin' of particle mass. (It seems that the idea of a massive particle is mysterious and problematic, while the notion of interacting fields is not.)

What is puzzling is how this slow sedimentation of ad hoc metaphysical assumptions became assimilated into physical lore and our laws of nature over the years. Young people entering physics today are already prejudiced and influenced by the inane nattering of the "Sunday scientists" and are in no position to critically examine the foundations of their own discipline.

To add an interaction term to our Dirac equation, we do not make the classical canonical electromagnetic replacement for the momentum, but simply include the term ieA. The four vector A represents the energy and momentum of the virtual photon or photons connecting the particle to the other particles involved.

Particles do not move through pre existing fields, no matter how attractive or convenient this picture may be. Particles, even test charges and experimental probes, are constantly interacting with everything and carry or drag these vertices around, constantly responding to and simultaneously altering them. Fields are a wonderful calculational device, a beautiful concept, but they are not real. They have no properties of their own.

Finally, let us consider our Dirac equation for a particle at rest. As we know, there is an uncertainty in the position and momentum, even for a particle 'at rest'. This conception is for a *point particle*. In our model, the electron is not a point particle, but a momentum vector tracing out a fundamental volume with a radius we characterize by the Compton wavelength. For a particle at rest, $\Delta x = \Delta y = 0$, while $\Delta z = \lambda_0 = h/m_0c$. Equation (14) becomes

$$H\,\psi = c\alpha \bullet \Delta p\,\psi \qquad (17)$$

$$\Delta p = h/\Delta z = h/\lambda_0 = m_0c \qquad (18)$$

$$H\,\psi = \alpha \bullet m_0c^2\,\psi \qquad ; \; \alpha = \alpha_z \qquad (19)$$

Even though we have removed the rest mass explicitly from the Dirac equation it still applies to a particle at rest as it seems it should. The rest mass of the electron is due solely to the rotational kinetic energy of the 'spinoring' of the angular momentum vector $L = \sqrt{3/2}\, h^{bar}$. The (sinusoidal) projection of the angular momentum of the electron along the z-axis is h/2. This allows the electron to absorb and emit energetic photons of angular momentum L = h.

We have proposed that the electron always spins to the left regardless of whether it is pointing up or down. Referring to Figure 1, we can see that the only conceivable difference between spin up and spin down would be a phase difference of $\pi/2$ in the relative rotation between any two angular momentum vectors. Nothing can point up all by itself.

Quantum mechanical electromagnetic induction

In our model, a mass current produces a gravitational magnetic field analogous to the magnetic field produced by a charge current. The original motivation for this perhaps not very original idea was to give the neutrino a magnetic moment of one sort or another. This proved fruitful in putting gravitation and electromagnetism on an equal footing and paved the way for our suppositions concerning the microscopic and macroscopic behavior of gravity.

(As an aside, we are no longer trying to force the neutrino and electron magnetic moments to have the 'same units', realizing that appropriate factors of α_G and α, or their roots, will solve the problem of the difference of charge between the two in any concrete calculations.)

Any effects of macroscopic mass currents are (usually) negligible in that they carry no net intrinsic angular momentum or spin. We remind the reader, again, it is the preferential spinning to the left or to the right of electrons and protons that make them stand out as electric charge.

In our original conception of the nature of electric charge, we assumed that there were certain 'resonances' in the frequency of spinning mass which were fixed, or quantized. The spinning electron mass induces a magnetic field which then induces an electric field which we experience as the electron magnetic moment and electric charge. (The muon mass would indicate the next resonance in this phenomenon of quantum mechanical electromagnetic induction, then the tau, and then the continuum produced by electric currents that we exploit in transformers and so on!)

Now, it is clear that the mass, or charge, and magnetic moment of a particle are two manifestations of one and the same phenomenon, namely spin angular momentum. It is just a matter of considering the different moments of mass. From afar, the particle appears as a point charge since the local oscillations of the angular momentum of the source charge have little effect at long distances. It is only 'up close' that this spinning becomes important as an effective time varying periodic electric charge.

So, an electron is not a spinning electric charge producing a magnetic field, but a spinning angular momentum that manifests as both. Neither electricity or magnetism are fundamental, as we know, but reflect a simpler underlying mechanism. You can't have one without the other.

The electric charge of the electron is fixed because the rest mass of the electron is fixed. The rest mass of the electron is fixed because it is! If the electron didn't have a fixed rest mass we wouldn't be here. One cannot argue with logic like that.

Particle families

In our formulation, electric charge is not an extra attribute of a particle above and beyond the angular momentum or mass of the particle. In fact, the electric charge can be considered strictly as a coupling constant denoting any particle with a *fixed minimum angular momentum*. Thus, it is not a mystery why the muon and the tau have the same charge as the electron, but only why their rest masses are what they are. The electron, muon, and tau *neutrinos* differ only in their initial lower frequency bounds, dictated by the charged particles' rest masses.

All three leptons are essentially the angular momentum vector $L = \sqrt{3}/2\ h^{bar}$. The three leptons only differ in the minimum rate of precession of this angular momentum vector. In keeping with our theme of the minimization of work, or the minimization of changes in kinetic energy during any interaction, as the foundational principle of physics, we decided that all interactions must minimize the difference between the rest masses and the kinetic energy of the particles involved, thus creating the need for families. Crudely put, at some point it is more expedient for nature to produce a slow muon rather than a fast electron.

Our model of muon decay is presented in Figure 2. The muon sheds its excess matter in the form of neutrinos. The 'massless' virtual electron is really just a glorified virtual photon since it *is no longer required to carry charge*. This is a new tweak to our model.
There is a factor of $(\alpha_W)^{1/2}$ at each vertex.

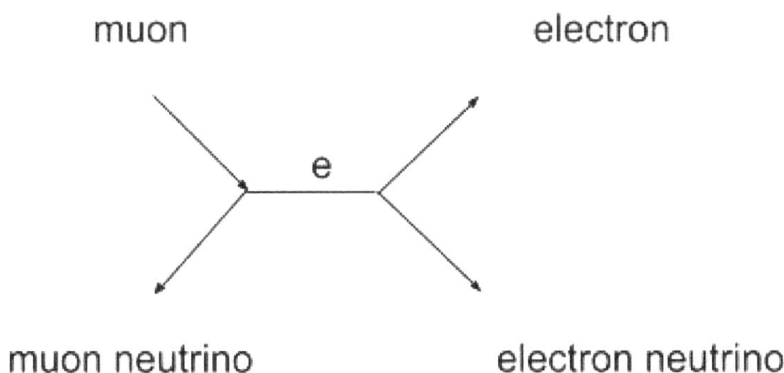

FIgure 2: Muon decay. The 'propagator' is virtually a virtual photon!

The leptonic table:

LEPTONS			ANTI-LEPTONS	
electron	electron neutrino	PARITY ⇔	electron antineutrino	positron
⇐	CHARGE	MASS ⇅	CHARGE	⇒
muon	muon neutrino	PARITY ⇔	muon antineutrino	anti-muon
⇐	CHARGE	MASS ⇅	CHARGE	⇒
tau	tau neutrino	PARITY ⇔	tau antineutrino	anti-tau
⇔	mass isospin	charge isospin ⇕	mass isospin	⇔

TABLE 1: The leptons and their interrelations; or the kleptogenesis of the leptoquarks.

Any lepton can be 'generated' from any other by the appropriate applications of the parity operator, the mass isospin operator, and our newly proposed 'charge isospin' operator. Mass isospin is the same as weak isospin in the standard model. The idea, of course, is that the neutrino and the electron have the 'same' mass and only differ in electric charge, and that the electron, muon, and tau have the same electric charge but only differ in mass. Particles and antiparticles differ only in parity; i.e. spin to the left or right !

Using various combinations of the step up and step down operators of SU(2) and SU(3), plus the 'parity operator', we can write any quantum mechanical interaction current in terms of the 'fundamental' neutrino neutral current.

Maybe nature knows about groups.

Time and space

In the beginning there was the universe. A universe much as we see now, with a similar distribution of mass and matter, stars and galaxies, intergalactic gas clouds and black holes. An eternal, boundless, flat and static, albeit self-renewing, universe where entropy need not apply.

One might say in the past, we've been "all over the map" concerning time and space. Perhaps. But we must remember we are dealing with essentially metaphysical concepts and trying to appropriate and quantify them to function in our physical theories and equations. In this sense the definitions of time and space are no different than the definitions of mass, force, and acceleration, except for the fact, of course, that acceleration assumes or implies time and space. Nothing can be fixed, or isolated, or defined without reference to something else.

This is essentially our theory of time and space. You need at least two objects or particles to define a space, plus some relative motion between the two to fix a velocity. A single particle plus an observer really won't do! Once you fix your observer "by definition" at the origin of your inertial coordinate system, it is only the relative separations, velocities, and accelerations of the particles that matter for physics. In any fixed coordinate system, time and space really do act as 'containers'; that is, *they do not act at all*. The space coordinates of any inertial coordinate system are three independent variables, while the *time* is merely a counter, dependent on some *relative motion in the coordinate system* thus defined. Time is a parameter, not a variable.

white space

Conclusion

We have no problem with the mathematics of quantum mechanics. Quantum mechanics fits quite nicely into and is in fact the foundation of our model. What we take issue with, and must vehemently protest, is the magical and mystical interpretations of quantum mechanics that have essentially become gospel for most people. Our mechanical model of the spinor accounts for the double valued nature of the electron wave function, the Heisenberg uncertainty principle, the Born rule; and qualitatively at least, the double slit experiment and quantum mechanical tunneling. All this is fairly clear in the case of a single free particle, at least. I believe the case for tunneling could be given a mathematical footing fairly easily.

In our model particles always have concrete values of spin angular momentum; both in magnitude and orientation. There is no spooky action at a distance. There is no wave function collapse. When an experimenter sets up an experiment, they do not create a wave function which they subsequently collapse. The idea is absurd. The wave function is a theoretical device used to calculate experimental outcomes. In our reconfiguration, the wave function is not a probability wave but rather represents the real time evolution or status of a particle's (or system's) energy, momentum, and angular momentum as the particle propagates or 'evolves' under the influence of mutual, equal and opposite, forces.

Our 'Equivalence Principle' is that everything is angular momentum; inertial mass, gravitational mass, electric charge, leptons, photons and … that's all there is! The question of particles possibly possessing two different types of mass always seemed to me more like a posturing than a serious scientific concern. What could this even mean? Perhaps it is a serious consideration for people eventually destined to also assign the lowly particle flavors, colors, and weak charge in addition to mass and electric charge!

In an attempt to 'mathematically' describe particle creation and destruction the metaphorical idea of the quantum mechanical vacuum spawned a metaphysical nightmare. The vacuum became more pervasive and more replete with convenient properties, even as it became ever more real in people's minds the more problems it was able to solve or resolve. A bubbling, magical cauldron indeed !

Particles are not created and destroyed. Particles merge, or combine, and then decay appropriately (i.e., in the most 'efficient' way) given the energetically possible available particles and the final energy-momentum phase space, and so on. If this process is best represented mathematically with creation and destruction operators then so be it, but *don't reify the vacuum.*

Nature abhors the quantum mechanical vacuum.

Books by Greg Feild

1. “A quantum mechanical theory of gravitational interactions”
   CreateSpace Independent Publishing, 8/29/2016

2. “Observations on the quantum mechanical nature of gravity”
   CreateSpace Independent Publishing, 10/8/2016

3. “On gravitation and electric charge”
   CreateSpace Independent Publishing, 10/29/2016

4. “On spin, mass, and charge”
   CreateSpace Independent Publishing, 11/29/2016

5. “On angular momentum, acceleration, and absolute motion”
   CreateSpace Independent Publishing, 1/1/2017

6. ”The Sinister Universe”
   CreateSpace Independent Publishing, 3/1/2017

7. ”On Parity and Isospin”
   CreateSpace Independent Publishing, 4/11/2017

8. “Reflections on the Sinister Universe”
   CreateSpace Independent Publishing, 5/12/2017

9. “On Current Physics”
   CreateSpace Independent Publishing, 6/11/2017

10. “A Critical Examination of Classical and Quantum Mechanical Waves”
    CreateSpace Independent Publishing, 6/18/2017

11. “On wave particle duality and the quantum of action”
    CreateSpace Independent Publishing, 7/6/2017

12. “On matter, mass, and motion”
    CreateSpace Independent Publishing, 9/14/2017

13. “On action and reaction”
    CreateSpace Independent Publishing, 9/24/2017

14. “A quantum mechanical theory of everything”
    CreateSpace Independent Publishing, 11/5/2017

15. “On Interaction”
CreateSpace Independent Publishing, 4/21/2018

16. “On Rotation”
CreateSpace Independent Publishing 8/19/2018

17. “Revenge of the Sinister Universe: The Reality of Everything’
CreateSpace Independent Publishing, 9/4/2018

18. “On Math, Physics, and Metaphysics”
CreateSpace Independent Publishing, 10/1/2018

19. “On Quantum Mechanics”
CreateSpace Independent Publishing, 10/15/2018

20. “On Epistemology and Ontology”
CreateSpace Independent Publishing, 10/21/2018

21. ”Toward a Metaphysics of Mass and Motion”
CreateSpace Independent Publishing, 10/29/2018

Compilations:

A. “The Universal Model of Our Sinister Universe: The First Ten Books”
CreateSpace Independent Publishing, 7/2/2017

B. “The Canons of the Sinister Universe:
The Last Four Books on the Universal Model of Our World”
CreateSpace Independent Publishing, 11/5/2017

C. “The Return of the Sinister Universe: The Immaculate Collection”
CreateSpace Independent Publishing, 9/4/2018

D. “The Battle for the Sinister Universe: The Heuristics”
CreateSpace Independent Publishing, 10/29/2018

22. “Postmodern Physics”
Kindle Direct Publishing, 10/28/2020

23. “Mechanism and Energetics”
Kindle Direct Publishing, 9/30/2021

Resources

Quantum Field Theory
Claude Itzykson, Jean-Bernard Zuber

Atomic and Quantum Physics
H. Haken, H.C. Wolf

Modern Elementary Particle Physics
Gordon Kane

Classical Dynamics of Particles and Systems
Jerry B. Marion

Foundations of Electromagnetic Theory
John R. Reitz, Frederick J. Milford, Robert W. Christy

Quantum Physics
Rolf G. Winter

Gauge Theories in Particle Physics
I. J. R. Aitchison and A. J. G. Hey

Quarks and Leptons: An Introductory Course in Modern Particle Physics
Francis Halzen, Alan D. Martin

Quantum Field Theory
F. Mandl, G. Shaw

Theoretical Mechanics of Particles and Continua
Alexander L. Fetter, John Dirk Walecka

The Theory of Spinors
Elie Cartan

Elementary Modern Physics
Richard T. Weidner, Robert L. Sells

Quantum Mechanics
Claude Cohen-Tannoudji, Bernard Diu, Franck Laloe

Philosophers

We have been hard on modern philosophers; accusing them of allowing and even aiding and abetting the physicists to run riot; riding roughshod over rhyme and reason.

In the book "Philosophical Aspects of Modern Science" (1932), C.E.M. Joad makes short shrift (in the most pleasant and politest way possible, of course) of the philosophical pretensions of Sir Arthur Eddington and Sir James Jeans. They had each proposed that Everything was Consciousness or Mind-Stuff. Were they properly chastised? Did anyone care? I don't know. The philosopher George Santayana was equally disdainful of these gentlemen's ideas

More recently, I've seen some modest potshots from some philosophers at the metaphysical posturings of physicists, but they seem wan and weary. Who can blame them? We need mutual cooperation, not one more 'culture war'. Plus, it's tiring and tedious to correct people.

A few more books:

"The Philosophy of Science", edited by David Papineau
Oxford University Press, 1996

"Philosophy and Mystification: A Reflection on Nonsense and Clarity", Guy Robinson
Fordham University Press, 2003

"The Investigation of the Physical World", G. Toraldo Di Francia
Cambridge University Press, 1981

"Philosophy and Spacetime Physics", Lawrence Sklar
University of California Press, 1985

Shop your local used book stores!

sarva asti

Empirical Enquiries

Greg Feild

June 5, 2022

About the author

Greg Feild is a former experimental high energy physicist.
He is not considered to be an expert in anything by anyone.

Buyer beware!

Abstract

In this book we examine some of the assumptions made in the formulation of modern physical models and theories, and question how valid or problematic these assumptions might be. Of course, many problems in fundamental high energy physics are well known and acknowledged, although apologists for renormalization and the like still abound.

A familiarity with some of my other works will be helpful to the reader, and is assumed in some places, although I will elaborate on some of *my own theories* here as well. If you are not already a fan, this book may not be for you!

Caveat emptor.

The quantum mechanical spinor

A brief recap. In our model an electron (or any lepton) has three linearly independent spin components; $s_x = s_y = s_z = h^{bar}/2$. An electron *is* the angular momentum vector $L = \sqrt{3}/2\ h^{bar}$. This vector 'precesses' (in a special spinor way) at a minimum rate determined by the electron rest mass. As the electron propagates and precesses, the projection of the angular momentum along the direction of travel varies sinusoidally (although the polarization remains constant!)

Our model of the electron spin and mode of propagation is depicted in Figure 1.

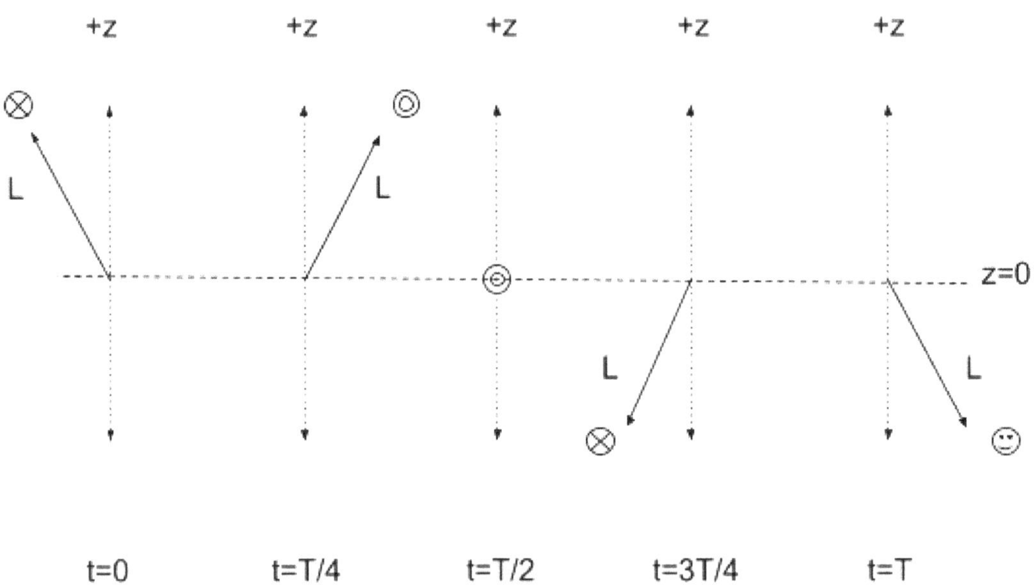

Figure 1: At rest with a lepton traveling in the z-direction. The spin angular momentum vector 'precesses' about the direction of motion, tracing out a closed, three dimensional figure eight. The x symbol represents motion into the page. The dot symbol represents motion out of the page. At time T/2, we see the angular momentum is *perpendicular* to the direction of travel. To represent an antilepton, simply swap the x symbols and the dot symbols.

Referring to Figure 1, we see that at time T/2 there is no projection of the angular momentum along the direction of travel, hence an electron is unable to absorb or *interact with a real photon* at these sorts of nodes. These nodes are where the particle catches up with all its virtual interactions and/or bounces off a container wall!

Introduction

Imagine a world where gravity and electromagnetism are the only forces, where mass is the only charge, and leptons and photons are the only particles. This is our model of the world.

In particular, it is the *angular momentum* or 'rotational' kinetic energy of a particle that constitutes the 'coupling charge'. (Energy and momentum are no longer independent concepts, as we know.)

Imagine a world where elementary particles are *massive* objects that *actually* spin.

More specifically, imagine a world where particles *are* spin. This is our world. Every aspect or property of an elementary particle can be reduced to or derived from spin. Spin determines *everything*; mass, energy, linear momentum, angular momentum, and *even* the electric charge.

In our model, interacting particles are constantly and continuously exchanging energy and momentum in a battle to win the upper hand and achieve their lowest possible relative mass-energy given the circumstances. The electric charge is merely *a constant* that enters into calculations to signify that the particle (e.g. an electron) has an intrinsic lower bound on its rate of spin; i.e., the particle rest mass.

Now, imagine a world where electrons spin to the left, and positrons spin to the right; a world consisting solely and completely of leptonic matter and 'antimatter', combined in equal proportions, comprising all the hadrons. Imagine a world without quarks or vacuums. This is *our* world.

Imagine a world where there are no forces or fields, only energy and momentum exchange; or action and reaction. The field concept is useful for solving many problems but fails miserably at describing the underlying 'nature of reality'. Carried to its 'logical' conclusion, field theory has charged particles interacting with themselves! In our model, particles do not interact with themselves. The very idea does violence to the concept of inter-action.

Imagine a world where the fundamental 'symmetry' is "for every action there is an equal and opposite reaction". A rotational symmetry. Imagine a world where there are no hidden symmetries, no broken symmetries, no confined and inherently non-observable particles, no collapsing wave functions, no ghostly waves.

Not so hard to imagine at all !

:)

On math and physics

All of the problems in modern physics today stem from the fact that people regard fields, wave functions, and the like as real physical entities rather than simple mathematical tools. This has been a disaster for physics and rational thought. Dramatic, but true.

The problem of an action at a distance between particles is supposedly solved by the introduction of fields. Now, however, each and every particle does not merely interact at a distance with other physical particles, but each particle creates a force field, *at a distance*, *everywhere*, at every point in space, regardless of the physical environment. This is particularly a problem for a classical static electric field, say. An electron 'at rest' has to maintain some constant force and/or energy field at each and every random point in space. We have a giant loss of 'efficiency' in our model with no real gain in explanatory power. How are these fields manifested and maintained *at a distance*? Of course, the modern answer is by virtual particle exchange, but in this model it seems there can be no 'force field' at a location where there is no particle (or measuring instrument!) which is one of the foundations of our model of particle interactions. It has been claimed that fields are real because they can be measured. I believe this is fallacious because measuring a field is not like measuring someone's height with a yardstick. You can't measure a field without introducing an interacting 'test charge'.

The main confusion in field theory today is the equating of 'internal' electric and magnetic fields representing the interactions between particles with the generation of 'external' electromagnetic radiation. Electromagnetic radiation is real and causes real action at a real distance, after the appropriate time delay, because it is composed of real particles called 'real' photons. It seems reasonable to assume that if 'external' interactions are caused by the emission, transmission, and absorption of 'real' photons, then 'internal' interactions should be caused by the "exchange" of 'virtual' photons. We agree with this modern conception up to a point. In our model, virtual photons are not emitted and absorbed (flitting about and fluctuating as they go) but instead represent the *constant* connections between *each and every* particle as conduits of energy and momentum exchange. A combination of these single interactions is often conveniently, and necessarily, represented by fields. However, when you eliminate the source(s) of the field, then the field itself must carry energy and momentum, as we know. Physicists often speak blithely of the idea of constant fields, without explicitly considering the work being done by the complex of batteries, electromagnets, etc, to sustain such fields. The electrons in the power sources are the actual sources and sinks for the energy and momentum of a particle under study in a particular electromagnetic field in the lab, *not the fields*.

Even in *classical* electrodynamics, the interaction of an electric charge with an electromagnetic field leads to the conception of a "renormalized mass", canonical momenta incorporating the fields, not to mention issues concerning cause and effect, *pre-acceleration*, etc.

In our model there are no fields, only particles exchanging energy and momentum constantly, and continuously, through dedicated lines called 'virtual' photons.

In electrodynamics, one usually begins 'classically' by solving Maxwell's equations *without* current sources, and quantum mechanically by quantizing the free classical electromagnetic fields. Unfortunately, this seems to be more than a merely pedantic procedure. The fields acquire energy and momentum, basically an existence, a life of their own, even before the introduction of sources and sinks. Basically, one has established a new sub-stratum for interactions; a Lorentz invariant aether, if you will. Is this an advance?

Even worse than the idea of fields carrying energy and momentum are the ideas of sourceless fields, and/or <u>constant fields</u> pervading the entire universe. It is not clear that these ideas are useful, even for 'thought experiments'. If fields and potentials prove to be unreal, then several particularly celebrated thought experiments lead to unfortunate (and incorrect) results.

Perturbation theory, a completely respectable approach to solving problems, seems corrupted when one can 'choose' to make a perturbative expansion in α, even if there is no 'classical' potential contributing to the perturbation, or even a correspondence of the 'perturbations' to the original set of eigenvectors of the original unperturbed Hamiltonian. We believe that given the 'correct' Lagrangian or Hamiltonian for a system, all higher order perturbative corrections will be in terms of the expansions of the relativistic mass and energy terms of the equations. Lagrangians which are formulated to yield the 'expected' or 'correct' equations of motion are also a little suspect. Maybe the 'target' equations are not ideal, but merely a prejudice, and a more rigorous Lagrangian might provide better equations of motion.

Then there is the celebrated Green's function devised for solving *any* potential problem, in principle, and responsible for the notorious propagators of quantum theory. Is the Green's function one of the most elegant, graceful, and beautiful mathematical formulations ever, or clunky, ugly, and hamfisted? The jury is still out.

Next, the vacuum; an inexhaustible bag of theory saving tricks! Historically, it seems the vacuum was invented by Dirac to accommodate the existence of positrons. If our model of matter and antimatter is correct, then positrons compose 50% of what we call ordinary matter. If so, then the vacuum is on shaky ground! One doesn't usually see these positrons because they are confined inside the protons and the neutrons. Unlike the quarks, however, these positrons (and electrons, and neutrinos!) can escape said confinement and be observed.

When an electron and a positron meet or collide, they do not annihilate one another in some new-fangled, magical fashion. The two spins *cancel* yielding a 'virtual' photon which will then 'decay' in the most kinematically favorable way possible. This model is a bit new-fangled, but it is not magical; sticking to classical concepts as best we can in this new domain.

Finally, gauge theory, the biggest offender of them all. By what authority should I "be able" to add an 'arbitrary' space and time dependent phase factor to my quantum mechanical wave function "without changing the physics", and when I cannot, by what right can I summon up a corresponding and compensating gauge field transformation to set things right?

When one makes a particular Lorentz transformation for an electrodynamic system, there will be a particular and corresponding change in the kinetic energy of the particles, and in the 'gauge' (~E and B) fields governing the particle interactions, but this is hardly 'arbitrary'.

In our model, the relativistic, gauge invariant, and Lorentz invariant wave function is

$$\psi = \exp(i(p \bullet x - (m-m_0)c^2t)/h^{bar}) \qquad (1)$$

where the rest mass term, m_0, represents the Lorentz invariant global phase factor and the relativistic mass term, m(x,t), represents the *local phase factor* corresponding to a given coordinate system or <u>chosen inertial reference frame</u>.

It is easy to see that this wave function satisfies the 'universal' Dirac equation

$$H \psi = ih^{bar}\partial\psi/\partial t = c\alpha \bullet p \, \psi \qquad (2)$$

It turns out Nature doesn't know about gauge invariance, but it does know about rest mass and relativistic mass. The term on the right hand side of equation (2) is, and must be interpreted as, the kinetic energy of the particle, an assertion we will examine in the next section.

From QED we have retained the idea of the virtual photon, assigning a certain reality to the four-vector A^μ which we have denied to the electric and magnetic fields! This is similar in approach to the standard model. However, in our model the four-vector A^μ represents the virtual photon 'wave function' connecting any two interacting particles (or a *single* particle interacting with an *ensemble* of particles) and not a set of independent fields. In other words, for 'internal' electromagnetic interactions between charged particles, the fields <u>cannot</u> represent 'free' virtual photons generated by electric "sources" <u>without the necessity</u> of any "sinks".

In other words; no test particle, no field; no measurement, no field; no free fields!

Electromagnetic radiation, or real photons, is generated by the acceleration of charges driven by external forces or influences. This radiation is emitted by the accelerated charges as individual photons, and <u>does not</u> represent an oscillation or disturbance of some already existing field established by the particle.

So, *virtual* photons represent the exchange of energy and momentum between particles acting *in concert*. Real photons represent particles *dumping* energy and momentum into the external environment. Both types of photons are involved in or can result in real interactions.

As seen, we also retain the idea of currents, particularly the idea of continuous electromagnetic (probability) conservation currents and the scattering currents of first order perturbation theory. A closer inspection of these currents reveal that they really, merely represent the linear momentum of a particle or stream of particles. (What else is there?) Any two individual moving particles, each constituting a current, are continuously interacting via one virtual photon. The energy and momentum of this one virtual photon are determined by the $1/R^2$ force law. Drawing inspiration from first order perturbation theory, we imagine that the ideal way to obtain the transition amplitude T_{if}, to all 'orders', for a scattering between two particles, would be by replacing the the first order 'initial state minus final state' current factor $p_A - p_B$, with a continuous integration over p from A to B, including, of course, both particles and the virtual photon. We wave our hands and sketch the mathematical representation as follows

$$T_{if} = \int_i^f J_1(r_1)\, 1/q^2\, J_2(r_2) \quad ; \quad q = p_1 - p_2 \qquad (3)$$

$$J_1(r_1) = p_1(r_1) \exp(-i\, p_1 \cdot r_1)$$

Neglecting coupling constants, etc., equation (3) seems to represent what is really physically happening when two "free" particles scatter off one another. The particles are constantly exchanging energy and momentum, each in order to obtain the lowest energy possible. This is why minimization of the action works so well. Particles don't feel forces and create no fields. Particles have no special quantum tags, or labels, not even the classical electric charge!

Energy states and energy differences are relative, local, and unique to each situation. This is why the minimization of the energy and energy conservation are trickier and more subtle concepts than they seem. Consider an electron 'at rest'. Certainly such an electron has less mass-energy, or kinetic energy, than an electron circling a proton at 1/137 the speed of light. Yet, we say that such a bound electron in a hydrogen atom has less energy than a free electron at rest! Such a statement makes sense only relative to the assumption of the Coulomb potential model. Any energy noted beyond rest mass energy is more of a bookkeeping value rather than an absolute, objective quantity. In absolute terms, two bound particles orbiting one another have more energy than the same two particles at rest. To claim the particles have formed some 'negative potential energy well' is a beautiful model but an absurd picture of reality. Unfortunately, the hydrogen atom is not formed by someone bringing an electron and a proton together 'by hand' from infinity, doing a certain fixed amount of work along the way. In our model, potential energy is a useful concept but is not considered to be a real or absolute quantity.

On spin, mass, and charge

In our model, spin, mass, and charge (and magnetic moment) are inseparable and inextricable manifestations of one underlying phenomenon; quantum angular momentum. The unfortunate, but inevitable, fact that they were uncovered or discovered separately makes putting them back together again all the harder!

Rather than a spinning (extended) sphere of mass and/or charge, *our* electron *is* spin. Our electron *is* the quantity of angular momentum $L=\sqrt{3}/2\ h^{bar}$ 'spinoring' about a preferred axis, or direction of travel, at a rate determined by the rest mass plus any additional relativistic kinetic energy. In spinoring, the angular momentum vector traces out an isotropic cone, 'flopping over' as it precesses about the preferred axis, circling twice before returning to its original position. This results in a time dependent, sinusoidal projection of the angular momentum, dL_z/dt, along the (preferred) z-axis, *which is the kinetic energy of the particle*!

Let us repeat. For a free particle,

$$dL_z/dt = h^{bar}\omega = mc^2 \quad (4)$$

For a particle *at rest*, the kinetic energy is

$$dL_z/dt = h^{bar}\omega_0 = m_0c^2 \quad (5)$$

The kinetic energy due to motion is given by the 'Dirac kinetic energy operator'

$$T\psi = c\alpha \bullet p\ \psi \quad (6)$$

We clearly see a factor of cp = cmv which has units of energy. We interpret the speed of light as the *speed of rotation of the angular momentum vector* **L**; c = |d**r**/dt|. The *frequency of oscillation of the angular momentum vector* is given by the linear momentum **p**. So the kinetic energy of an elementary particle is completely 'mechanical' ! For a two component spinor traveling along the z-axis we can extract the relation

$$cmv = mc^2 - m_0c^2 \quad (7)$$

and

$$-cmv = mc^2 - m_0c^2 \quad (8)$$

So, kinetic energy can be positive or negative! It is all a part of the conservation of energy.

In our model, mass is the coupling charge, and for electromagnetic interactions we replace the electric charge, e, with $e(m/m_0)$. This leads to the following expression for the electron magnetic moment

$$\mu = (e/m_e)(h^{bar}/2c)(1 + \tfrac{1}{2}\, v^2/c^2 + \tfrac{3}{8}\, v^4/c^4 + \dots) \qquad (9)$$

or

$$\mu = \mu_0\, (1 + \tfrac{1}{2}\, v^2/c^2 + \tfrac{3}{8}\, v^4/c^4 + \dots) \qquad (10)$$

For an electron 'at rest' we will assume the velocity, v, to be Δv, the uncertainty in the particle velocity. If we also assume, explicitly

$$\Delta x \Delta p \equiv h^{bar}/2 \qquad (11)$$

and take the uncertainty Δx to be equivalent to the Compton wavelength (see Figure 1)

$$\Delta x = \lambda_0 = h/m_0 c$$

$$\Delta p = h^{bar}/2\lambda_0$$

$$\Delta v = \Delta p/m_0 = h^{bar}/2\lambda_0 m_0 = c / 4\pi$$

So we can now write, for the electron at rest,

$$(\Delta v/c) = 1 / 4\pi$$

and

$$\mu = \mu_0\, (1 + \tfrac{1}{2}\, (1/4\pi)^2 + \tfrac{3}{8}\, (1/4\pi)^4 + \dots) \qquad (12)$$

This is the best that our model can do to accommodate any observed anomalous magnetic moment for the electron 'at rest'. Admittedly, the exact choice of the uncertainty relationship equation (11) is arbitrary. This model should work better for measurements made on electrons 'in flight', with no need for experimenters to correct back or extrapolate to a 'rest mass' value.

If our result, equation (12), does not agree with current measurements, is our model dead? Of course not! There is still a possible contribution from our 'gravitational magnetic moment' to consider. In addition, we may well ask how many assumptions from the standard model 'contribute' to the current measurement of the electron magnetic moment.

The proton

Imagine if you will, if you can suspend all disbelief for just a moment, a proton consisting of two positrons and one electron. These three leptons will be moving very fast and will be very close together. They will be 'very' massive. The quantity $1/R^2$ will be large. The subsequent balance between the *gravitational* and electromagnetic forces will act as a constant potential and hold them together, forever.

In our latest model of the proton, the two electrons now form a closed shell around the positron 'nucleus'. In addition, our model of the electron allows us to understand the nature of all such closed shells. One electron must have spin up while the other must have spin down. In our model, the difference between spin up and spin down is solely due to a phase difference of $\pi/2$ in the electron spinor wave function. This *same phase factor* also accounts for the difference in helicity between the very same two electrons. Thus, two electrons can share the same orbital. They are 180 degrees out of phase, but still spinning in the same direction.

Incidentally, this spin up and down of which we speak, is actually the *helicity of the electron*, or the projection of the electron momentum 'along the direction of travel' (if you will allow the reintroduction of such a crude concept!). This is a *different*, disticint, and separate spin up or down from that which would be exhibited if said bound electrons were to be described in an internal or external magnetic field, orthogonal to the direction of travel. To reiterate, the two spin degrees of freedom which allow two electrons to share the same orbital via the Pauli exclusion principle (i.e. positive or negative 'helicity'), is not the same as the two spin degrees of freedom that allow the same electrons to point up or down in a magnetic field. If we associate s_x with the helicity and s_z with the spin up or down, then with these two determinations will also allow the determination of s_y as well, via the familiar spin commutation relations. There can at most be two independent components of the electron spin because the absolute value of the total spin must always be $S = \sqrt{3}/2\ h^{bar}$. It seems this is why complex numbers enter into Pauli spin matrices; the system is *mathematically* undetermined and the component s_y can swing either way! (This is the same reason why complex numbers creep into energy and momentum relationships. Energy and momentum are not independent, at least not like in the good old days, when wiser men than we argued about which should be considered the more primary quantity of motion.)

However, once the three components of the spin are fixed *physically*, then they are fixed, whether we can predict all three components or not. This has been proved **repeatedly** in 'quantum entanglement experiments'. The fact that certain quantities are underdetermined does not mean they are undetermined. Why people would disavow this logical fact in favor of spooky-action-at-a-distance is one mind boggling mystery.

But, back to our model of the proton.

We imagine the 'ground state' of the proton to be the strong force analogue of an helium atom, with a positron nucleus substituted for the proton nucleus, and α_s substituted for α.

What if the world were the 'planetary model' 'all the way down' ?

We have derived the strong coupling constant (14) to be

$$\alpha_S = (G/4\pi\varepsilon)^{1/2} (2m_e e/h^{bar}c) (1 + \tfrac{1}{2} v^2/c^2 + \tfrac{3}{8} v^4/c^4 + \ldots) \qquad (13)$$

This is cool but not covariant … but a good start we hope. A question arises: if this model is legitimate, would we consider the rest mass of the proton to be two or three times the relativistic kinetic energy of an electron confined to the radius of a proton in a hand waving calculation? In other words, would the positron be considered to be relatively at rest, as the proton is in the helium atom model?

Our previous model of the proton invoked one dimensional coupled harmonic oscillations of the two electrons and the one positron. Perhaps this could represent excited states. Such harmonic oscillator states could also rotate as a whole, etc.

In our 'dedicated virtual photon model' of particle interaction, the number of virtual photons necessary for the description of the interaction between a given number of particles follows Pascal's pyramid scheme! In other words: one particle, zero photons; two particles, one photon; three particles, three photons; four particles, six photons; five particles, nine photons; etc. !

In this model the proton would be considered to be three particles connected by three continuous, conditioned and *correlated* virtual photons rather than three independent particles producing three independent fields.

The Takeaway ? Quantum Gravity holds the proton together.

Classical Mechanics versus Quantum Mechanics

Quantum mechanics is not as different from classical mechanics as it may seem. Energy and momentum are conserved. Our ordinary definitions of space and time, our classical coordinate systems, and the differential and integral calculus are all applicable and appropriate for use in describing the world at the smallest distance scales probed to date. The classical Coulomb potential is the correct starting point for constructing the Schrodinger equation for the hydrogen atom. Subatomic particles spin. Macroscopic objects spin! Spin should have been expected? Spin should have never been relegated to a quantum 'label'.

We are not trying to minimize many of the interesting and unique features of quantum physics and quantum mechanical systems. However we believe they are not as weird as they seem and can be described by physical models derived from classical physics, with the appropriate quantum mechanical tweeks, of course! Our physical models of photon and lepton spin are the best examples. We believe these spin models account for all the peculiar 'quantum mechanical' behaviors of subatomic particles from tunneling to diffraction.

Besides quantization itself (which should no longer be a surprise) and the uncertainty principle, the third great revelation from the investigation of the microscopic realm is the existence of quantum spin statistics which themselves are derived from particular mechanical models of particle energies and their spins. (We shan't discuss spin statistics here!)

In both classical mechanics and quantum mechanics there are three canonical, conserved quantities for closed systems: E, L, L_z. These values; the energy and two values of the angular momentum, are constants of the motion. These facts alone should be considered just as amazing as anything else about the microscopic world. From the electron to the hydrogen atom to our planetary system to spinning galaxies, the conservation of these three quantities represents a Unity across distance scales that should brace the mind.

In classical mechanics we can also predict the position and velocities of objects while in quantum mechanics we can only predict averages. However, in classical mechanics one starts with position and velocity measurements and then derives the energy and angular momentum of the system from the equations of motion. In quantum mechanics, one starts with energy and angular momentum measurements of a system, and *at best* can predict average values for particle positions and velocities from the corresponding wave functions. These values are undetermined by construction. Position and velocity cannot be determined absolutely at any time and so cannot be the foundation for the (wave) equations of motion. If a theory could yield definite values for the position and velocity of subatomic particles inside a hydrogen atom, for example, we could not verify these values in any case. However, I do believe in these definite values. To claim the proton is a particle and the electron is a probability cloud is sizeism, pure and simple. They are both very small.

Just because the value of a quantity is underdetermined in our mathematical formulations does not mean that it is undetermined in reality. People would rather believe in nonlocality than believe that all quantum variables have distinct and determined values at all times. Why?

The same people who would caution one not to consider the electron a spinning ball of mass and/or charge (it's not) or even a particle at all, wholeheartedly embrace the ideas of the *physical* superposition of orthonormal quantum wave functions describing an unknown quantum state, *physical* wave function collapse, decoherence, entanglement, etc.

Unification

In the standard model, the 'three forces' are thought to merge at some ginormous energy scale where the various asymmetries, hidden, broken, or otherwise, which seem to generate the appearance of the different, or separate forces, no longer hold sway. We believe this is the wrong approach for several reasons; the main reason being that we do not believe this is the way that the forces or their relative strengths are generated or realized !

To our mind, the three (to four) forces do not merge in strength and efficacy at some absolute energy scale. Rather, we picture the four different forces as different manifestations of *only one* (or two !) forces (gravity and electromagnetism) acting at widely varying, and wildly different velocities and particle separations; and thus different energy 'scales'.

In our model, it is the relative masses, velocities, and separations of particles *alone* that determine the strength of the force between them (plus a multiplicative universal coupling constant or two, of course). *Angular momentum is the fundamental and only 'charge'.*

More precisely and less succinctly, all interaction between particles is due to the absolute difference between the sums of their internal and intra-particle angular momentums. The internal angular momentum determines a particle's mass, and the relative angular momentum between particles is given by their relative velocities and separations. This *sum* of these various angular momenta of a 'closed system' is invariant, and this sum, or angular momentum balance or imbalance, will determine the future evolution of a system. Particles are constantly trading angular momentum and/or mass-energy in the eternal quest for balance, equilibrium, and rest.

Profundity

We see, then, that both in the case of the indeterminacy principle and in that of von Neumann's theorem, conclusions have been drawn concerning the need to renounce causality, continuity, and the objective reality of individual micro-objects, which follow neither from the experimental facts underlying the quantum mechanics nor from the mathematical equations in terms of which the theory is expressed.

– David Bohm
Causality and Chance in Modern Physics

Including an epigram from David Bohm is a bit disingenuous, of course, since we reject his ideas of hidden variables, sub-quantum mechanical processes, or some reality 'deeper' than quantum mechanics. However, his critique is cogent.

Physicists' perpetual use of the word deep is both confounding and 'concerning'. Do they mean 'deep' as in submicroscopic, or 'deep' as profound? The mistaken idea that we can probe ever shorter physical dimensions or delve ever deeper into physical processes or aspects of reality is driving physics to ruin. Matter is not infinitely divisible, nor is it some localized stable configuration of a multitude of infinite, interacting, fluctuating fields; ones we may some day further divine. The first idea, of dividing matter indefinitely seems plausible at least, since space can in principle be so divided. The second idea seems fanciful at best. (Protons *never* decay.)

'We now know' that mass and matter are discrete, while space is continuous. Particles, both photons (mass) and leptons (matter), are well defined instances of angular momentum, propagating through space and passing through each and every physical point, as they traverse a well defined path. Space contains an infinity of points, and in principle a particle could be at any one of these points at any particular time. Matter, however, is not infinitely divisible. Matter is composed of *quantized units of fundamental angular momentum*. We consider the simplest elementary particles (photons, neutrinos, electrons) to be *oscillatory* instances of this angular momentum.

The uncertainty in the position and momentum of an elementary matter particle (one of Mr. Bohm's main concerns) is due to the very oscillatory nature of the particle itself. Rather than bouncing around randomly any uncertainty in a particle's position is due the effective radius or wavelength of the particle, which makes it a bit of a moving target for precise measurements.

I see no advantage to calling physics deep. It is an empty word.

Physics is fun. Let's leave profundity to the poets!

Conclusion

Physicists lack imagination; creative imagination at least.

The basic building blocks of nature are 'quantized' units of angular momentum. These units can only be combined in certain ways and change their configuration in certain ways determined by a fairly simple set of rules. This was discovered by scientists, to their 'shock and dismay', nearly one hundred years ago. In retrospect, we see they should have been delighted. The world is made up of bits! However, they had just wrapped their brains around free-floating fields; electricity and heat were still fluids, or still fluid in their conception, and all charge, all matter, was smooth and continuous.

Unfortunately, the concept of quantization was not enough to kill off the concept of the field. Most likely, the hypostatization of the wave function was partly to blame!

The total energy and momentum of an electromagnetic wave is equal to the sum of the energy and momentum of the individual photons in said 'wave'.

How could such a wave be considered anything more than a continuous set of wave fronts of oscillating photons?

Why wouldn't an individual oscillating photon not 'diffract' when it encounters an obstacle of two slits separated by a distance comparable to the 'radius' of the oscillating photon?

As long as people believe, and are educated to believe, that waves, wave functions, and fields are real things, no real progress can be made in physics. There will be technological progress, of course, but no true ontological or epistemological understanding. During the iron age, iron, copper, bronze, tin, silver, gold, were mined, smelted, and alloyed, but people still thought the four elements were earth, wind, fire, and water. Technological progress will always outpace theoretical speculations and explanations.

Most of my books are shorter than many people's blog posts. Perhaps they should be blog posts! However, I could not entrust even my simplest thoughts to anything as ephemeral as the internet. I humbly provide some food for thought and let the reader think for themselves, do some of the relevant, and much needed hard work, and maybe come up with something nice, new, and surprising of their own that they will feel compelled to communicate to the wider world with more enthusiasm than I can muster. Much more, I hope.

About the author

I earned a Ph.D in experimental high energy physics from the Pennsylvania State University working on HERA at DESY in Hamburg, Germany studying photoproduction and deep inelastic scattering in electron-proton collisions.

I did my postdoctoral studies with Yale University working at Fermilab on the CDF experiment at the Tevatron. My primary research interest was particle hadronization in quarkonium production in proton-antiproton collisions.

Books by Greg Feild

1. “A quantum mechanical theory of gravitational interactions”
   CreateSpace Independent Publishing, 8/29/2016

2. “Observations on the quantum mechanical nature of gravity”
   CreateSpace Independent Publishing, 10/8/2016

3. “On gravitation and electric charge”
   CreateSpace Independent Publishing, 10/29/2016

4. “On spin, mass, and charge”
   CreateSpace Independent Publishing, 11/29/2016

5. “On angular momentum, acceleration, and absolute motion”
   CreateSpace Independent Publishing, 1/1/2017

6. ”The Sinister Universe”
   CreateSpace Independent Publishing, 3/1/2017

7. ”On Parity and Isospin”
   CreateSpace Independent Publishing, 4/11/2017

8. “Reflections on the Sinister Universe”
   CreateSpace Independent Publishing, 5/12/2017

9. “On Current Physics”
   CreateSpace Independent Publishing, 6/11/2017

10. “A Critical Examination of Classical and Quantum Mechanical Waves”
    CreateSpace Independent Publishing, 6/18/2017

11. “On wave particle duality and the quantum of action”
    CreateSpace Independent Publishing, 7/6/2017

12. “On matter, mass, and motion”
    CreateSpace Independent Publishing, 9/14/2017

13. “On action and reaction”
    CreateSpace Independent Publishing, 9/24/2017

14. “A quantum mechanical theory of everything”
    CreateSpace Independent Publishing, 11/5/2017

15. “On Interaction”
CreateSpace Independent Publishing, 4/21/2018

16. “On Rotation”
CreateSpace Independent Publishing 8/19/2018

17. “Revenge of the Sinister Universe: The Reality of Everything’
CreateSpace Independent Publishing, 9/4/2018

18. “On Math, Physics, and Metaphysics”
CreateSpace Independent Publishing, 10/1/2018

19. “On Quantum Mechanics”
CreateSpace Independent Publishing, 10/15/2018

20. “On Epistemology and Ontology”
CreateSpace Independent Publishing, 10/21/2018

21. ”Toward a Metaphysics of Mass and Motion”
CreateSpace Independent Publishing, 10/29/2018

Compilations:

A. “The Universal Model of Our Sinister Universe: The First Ten Books”
CreateSpace Independent Publishing, 7/2/2017

B. “The Canons of the Sinister Universe:
The Last Four Books on the Universal Model of Our World”
CreateSpace Independent Publishing, 11/5/2017

C. “The Return of the Sinister Universe: The Immaculate Collection”
CreateSpace Independent Publishing, 9/4/2018

D. “The Battle for the Sinister Universe: The Heuristics”
CreateSpace Independent Publishing, 10/29/2018

22. “Postmodern Physics”
Kindle Direct Publishing, 10/28/2020

23. “Mechanism and Energetics”
Kindle Direct Publishing, 9/30/2021

24. “Deconstructing Modern Physics”
Kindle Direct Publishing 1/26/2022

Resources

Quantum Field Theory
Claude Itzykson, Jean-Bernard Zuber

Atomic and Quantum Physics
H. Haken, H.C. Wolf

Modern Elementary Particle Physics
Gordon Kane

Classical Dynamics of Particles and Systems
Jerry B. Marion

Foundations of Electromagnetic Theory
John R. Reitz, Frederick J. Milford, Robert W. Christy

Quantum Physics
Rolf G. Winter

Gauge Theories in Particle Physics
I. J. R. Aitchison and A. J. G. Hey

Quarks and Leptons: An Introductory Course in Modern Particle Physics
Francis Halzen, Alan D. Martin

Quantum Field Theory
F. Mandl, G. Shaw

Theoretical Mechanics of Particles and Continua
Alexander L. Fetter, John Dirk Walecka

The Theory of Spinors
Elie Cartan

Elementary Modern Physics
Richard T. Weidner, Robert L. Sells

Quantum Mechanics
Claude Cohen-Tannoudji, Bernard Diu, Franck Laloe

“The Philosophy of Science”, edited by David Papineau
Oxford University Press, 1996

“Philosophy and Mystification: A Reflection on Nonsense and Clarity”, Guy Robinson
Fordham University Press, 2003

“The Investigation of the Physical World”, G. Toraldo Di Francia
Cambridge University Press, 1981

“Philosophy and Spacetime Physics”, Lawrence Sklar
University of California Press, 1985

“Inconsistency, Asymmetry, and Non-Locality: A Philosophical Investigation of Classical Electrodynamics”, Mathias Frisch
Oxford University Press, 2005

a chapbook

imagine a world where it pays to keep physics weird

www.ingramcontent.com/pod-product-compliance
Lightning Source LLC
LaVergne TN
LVHW080553160826
845677LV00010B/1830

* 9 7 9 8 8 4 8 3 6 8 1 4 7 *